CONTENTS

ACKNOWLEDGEMENTS

The author and publisher are grateful to the following for permission to reproduce material.

British Telecom for the Weatherdial advertisement on page 129.

Dundee University for the weather satellite images within pages 135–83.

European Space Agency for the cover photograph.

The Independent for the five weather maps on pages 42–3.

Meteorological Office and the Controller of Her Majesty's Stationery Office for the weather maps on pages 30, 100–7, 131–4.

Ordnance Survey and the Controller of Her Majesty's Stationery Office for the map on page 80.

Telephone Information Services for the Weathercall advertisement on page 128.

Western Mail, Cardiff, for the newspaper articles on pages 34–5.

The author and publisher are also grateful to Ross Reynolds, Department of Meteorology, University of Reading, for his advice on the selection of case study images.

NOTES FOR TEACHERS

A Introduction

This resource and activity pack has been designed to be used with students following a Geography GCSE or Standard Grade course. Most of the resources and activities can also be adapted for both younger and older students. In the main, they are suitable for use with a wide range of abilities.

The aim of the pack is to integrate the use of live satellite images from Meteosat or NOAA satellites (see sections C–E below) into a study of weather. The use of live images means that students can actually see the weather as it happens, giving an immediacy and relevance to the study of weather which historic images, even if only a few days old, can never provide. However, the pack includes a set of high-quality satellite images, and many of the activities can be done using these rather than live images. In this way the pack can also be used by students who do not have easy access to satellite receiving equipment.

Although most of the major concepts, skills and attitudes associated with the subject of weather are covered to sufficient depth for GCSE and Standard Grade, the resources are not intended to make up a complete teaching pack on weather. Rather, it is hoped that teachers who have weather satellite receiving equipment of whatever type, together with those who do not, will be able to use the resources to enable students to look at and interpret satellite images for themselves in the context of their GCSE or Standard Grade work. Consequently, some of the activities will be of relevance to a particular course, and others not.

The pack is intended to be used in an active, investigative way. Often, a teacher is tempted to talk about a satellite image, either in a book, or on a slide, or even as a live image on a screen. It is hoped that the activities in this pack will allow the students to take on the role of weather interpreters and in this way have direct experience of using satellite technology for themselves.

The pack focuses specifically on the weather in the UK and Europe – to have covered global weather and climate in a pack of this nature would not have been practicable. However, all the satellite image interpretation techniques can obviously be used to help understand any satellite images, and students should be encouraged, if the school has a suitable receiving system, to look at images covering as many parts of the world as possible.

The pack also has a background section on the way satellites work (see section C below). For practical reasons associated with the length of the pack, this information has been written for teachers, rather than as activity material for students. Although not strictly geographical, information like this is important because students often ask questions about the technology involved in getting live images into the classroom.

Finally, each section of work is followed by an objectives checklist. It is useful for the students to have copies of these so that they can be filled in as they progress through the activities. The lists should provide them with a positive statement of what they know, understand and can do, and should help them to clarify exactly why they are carrying out each activity.

There are notes on specific sections and appendices at the end of these general notes (see section F below). They highlight activities that may be difficult in schools with more basic receiving equipment, and give background information and suggestions for getting the most out of the material.

B How to Use this Pack

The activities in the pack all follow a similar format, with an introductory section and an activities section. They are designed to be photocopied on to A4 paper or card and given to the students. Most are two or three pages long.

Although the activities in this pack are arranged in a logical order in terms of the study of weather, there is no need to start at the beginning and work through every activity that has been provided. Some may be more appropriate to a particular course than others.

When using any of the activities, however, the following points should be borne in mind:

1 It is assumed that students are using monochrome and not colour images. This is because students find these images easier to work with as there is an obvious gradation of shading, say from white through to black, which is immediately understood. Some of the enhanced colour shadings used often make interpretation difficult.

There are also other, cost reasons for using monochrome images. To reproduce colour images in a pack of this nature would have been expensive. In addition, if teachers are able to make hard copies of images, either on a video printer or a dot-matrix printer, then these will probably be in black and white. Not all teachers will have access to a colour monitor either.

2 Five high-quality copies of each of ten different satellite images are provided with this pack, in Appendix 6. These image sets have two intended uses.

Firstly, each of the images gives an example of a good case study observation of a particular weather condition. Used in conjunction with the appropriate annotated weather maps found in the activities, they provide a useful guide to the interpretation of satellite images – a kind of satellite image key which students should find helpful when looking at live images of their own.

Secondly, the image sets widen the scope of this pack as a useful resource for those schools which perhaps only possess a basic receiving system, or have not even invested in one as yet. A number of activities can be used with the images in the image set, and these are marked with the symbol ✥. In addition, Appendix 5 contains the weather maps associated with each image so that teachers can construct their own activities based on the weather situations represented by the images. (*Note:* the images are the copyright of the University of Dundee and may not be reproduced in any way without the copyright holder's written permission.)

3 Exactly how live images are used in the classroom will depend to a large extent on the technology available within the school. If a school has no facility for making a hard copy of an image, the first step in most of the activities will be to transfer relevant information from the image to a base map. Obviously this means that only part of a class will be able to work around the monitor screen at any one time. Several classroom management issues arise from this, and to help deal with these, many sections of the pack contain additional supporting activities which do not need a satellite image to complete. These are marked with the symbol ◈.

Such issues do not arise for those schools which have the facility to make a hard copy of an image. In these situations students can work directly on the image itself, giving a greater degree of accuracy. It is suggested, however, that for each image students are encouraged to identify clearly which part of the globe they are looking at!

4 The objectives checklists that follow each section are an essential part of the pack. It is suggested that students are provided with copies of the appropriate checklists, depending on the activities they have completed, and that after each activity, or perhaps at the end of a series of units, the activities are reviewed and the checklists completed as a record of achievement. The checklists could also be used before an activity is attempted so that the student is given some idea of what he or she is expected to achieve.

5 Most of the activities in the pack have been pitched at a middle level of satellite technology. Schools with the more basic receiving equipment will find a degree of difficulty in carrying out some of the activities. So that frustration and disappointment do not occur, these activities have been highlighted in section F below.

On the other hand, schools which possess more sophisticated receiving equipment might feel that the activities do not use the full potential their resources offer. They are encouraged to go further and experiment with techniques such as image processing and the like. The activities in this pack should be a gateway to further exploration, not a barrier.

6 It is hoped that the activities in the pack are stimulating and accessible for all levels of ability – they have been designed with this in mind. They are also relatively flexible, and can be used by the imaginative teacher in a wide variety of ways. In particular, there are a large number of coursework ideas in the pack, since the study of the weather in a local area is an ideal coursework or assignment topic. Many of the activities not specifically suggested as coursework ideas can easily be adapted to suit coursework requirements.

Finally, the pack also includes a number of appendices. A skills matrix has been provided, for use if there is a need to target particular skills in an activity. Base maps have been drawn for the Meteosat C2 and C3 sectors, for each of the D sectors, and for the whole-earth images, as well as for images provided by GOES East. A basic NOAA outline map has also been drawn. These the students should find helpful for making copies of images used in the activities.

The image sets and the relevant Meteorological Office weather maps are given in the appendices too, and form an invaluable teaching resource.

The appendices also contain two 'bibliographies' – one is a list of useful addresses concerning satellites and satellite resources, and another a list of weather maps and data that are easily available in the media.

The final two appendices contain teaching material referred to in the activities – some weather map symbols that could be used on the large plastic wall maps available from the publishers or on school-produced weather display maps, and a depression model.

C About Weather Satellites

Students are generally highly inquisitive – especially when it comes to high technology. Although perhaps they tend to accept satellites and satellite images as a normal part of life in the late twentieth century, when they actually watch a live image being received in the classroom their curiosity and interest are aroused. As a result, they ask questions, and the aim of this section is to provide teachers with some background information about satellites which, hopefully, will help them to answer most of the points students will raise.

There are many well-known remote sensing satellites too. The NOAA (National Oceanographic and Atmospheric Administration) satellites and Meteosat are weather satellites, and the images they produce can be seen each day on television weather forecasts. Landsat, on the other hand, collects information about land use, and provides high-quality land-use images for large parts of the earth. The new French SPOT satellite does much the same work as Landsat, but has a higher resolution and so provides a higher quality image.

Different kinds of satellite

There are basically two different kinds of satellite. There are *communications satellites*, which simply act as giant reflectors in the sky bouncing data from one place to another, and there are *remote sensing satellites*, which collect special information about the earth's surface and atmosphere to help us understand how the earth works. The diagrams below illustrate the differences between these two groups.

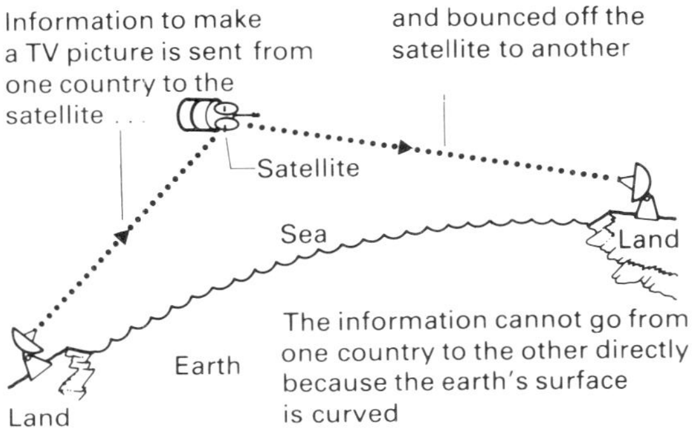

A communications satellite sending a TV picture

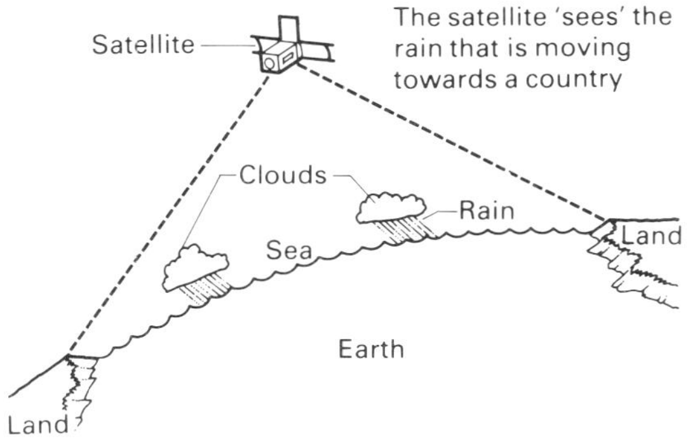

A remote sensing satellite collecting weather information

Intelsat is a good example of a communications satellite. It is responsible for sending thousands of telephone conversations from one country to another each day. Communications satellites are also used to beam television pictures across the globe.

How does a satellite work?

Most satellites are about the size of an average estate car, and are launched into orbit using either a rocket such as the European *Arianne*, or the USA's *Challenger* space shuttle. The *orbit* of a satellite is simply its path as it goes round the earth. The orbit can be as low as about 600 km in the case of Landsat, or as high as 36 000 km in the case of Meteosat and other geostationary satellites. (See the section below on different kinds of orbit.)

Satellites are usually powered by solar energy, collected by solar panels attached to the body of the satellite and looking like giant wings. Sometimes satellites are powered by nuclear energy, as was the Soviet one that crashed to earth over Canada. This event was unusual, because although satellites do not last for ever, they normally either stay in space, or burn up as they re-enter the earth's atmosphere. It has been estimated that there are about 7000 people-made objects in space at present, and of these fewer than one in twenty are actually working satellites. The remainder consist of old rocket tanks, dead satellites and other debris.

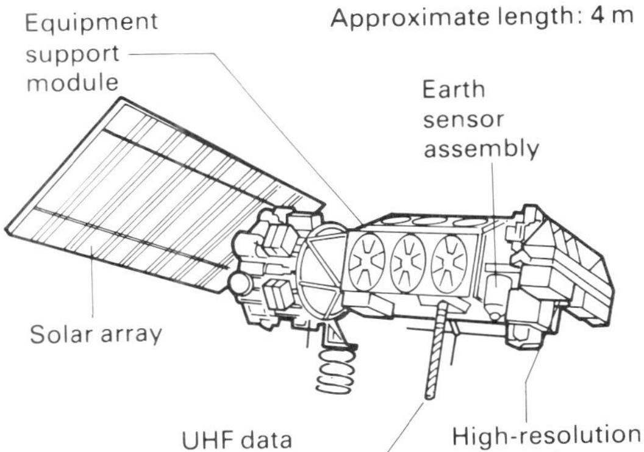

NOAA

To receive messages or information from a satellite an aerial is needed, together with a receiver and some means of decoding the data sent down by the satellite. In the case of Meteosat, a dish aerial has to be pointing directly towards the satellite to pick up its signal. The place on earth where the aerial and receiver are located is called a *ground station*, and if your school has a satellite receiving station, then it is as much a ground station as, say, Madeley in Herefordshire – one of British Telecom's major satellite receiving stations.

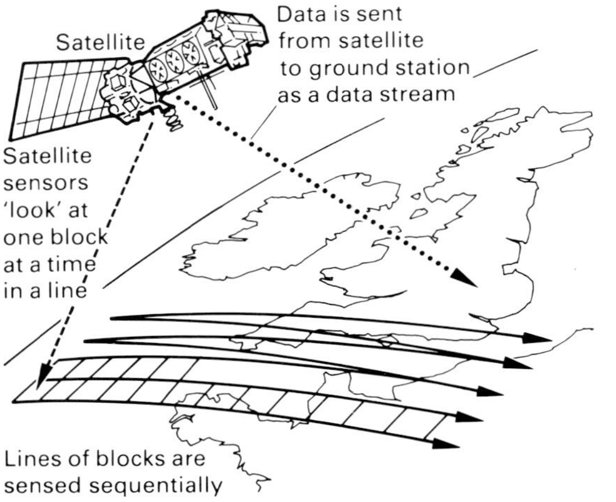

A satellite scanning an area

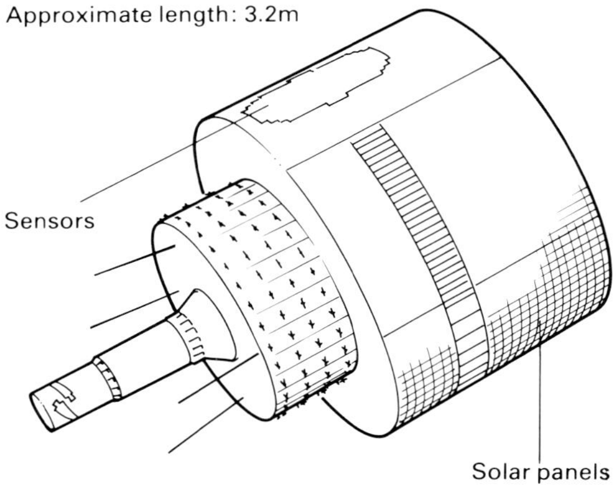

Meteosat

Unless you have access to information from some of the spy satellites in space, the image you will receive will not be a photograph. Remote sensing satellites have an array of sensors on them which have the job of picking up information from the earth's surface, usually in the form of reflected light or bands of radiation.

A satellite image will differ from a photographic image in another way as well. Usually, the satellite will 'see' or sense the earth's surface in a series of blocks – in other words, it does not sense absolutely everything on the surface of the earth. Landsat, for example, looks at blocks which are 30 square metres, whilst the new French SPOT satellite has a better resolution of 10 square metres. Information about each block forms a single picture element (or *pixel*) on the finished image. Instead of 'seeing' a large area all at the same time, as a camera would, a satellite scans an area line by line, with each line being made up of a number of blocks (or pixels). The next diagram shows this process. Information about each block – say, in the case of a meteorological satellite, infra-red radiation – is measured and coded as a number. A set of numbers about each line scanned is sent down to the satellite receiver in the form of a *data stream*, and these numbers then have to be decoded and turned back into an image again. This is the reason why, when you look at an image being received and appearing on a monitor sceen, it will build up line by line rather than becoming visible all at once.

Different kinds of orbit

Satellites go round the earth in a number of different ways.

The *polar orbiting satellites*, such as NOAA, SPOT and Landsat, are in a fixed orbit roughly from pole to pole. The earth spins beneath them. Because of their relatively low orbit height – anything from 600 km up to 1000 km – they cannot sense more than part of the earth's surface at any one time; but in a given period, because of the rotation of the earth, the satellite will pass over the whole earth's surface. This period varies, of course, depending on the speed that the satellite is travelling at, but in the case of Landsat it is sixteen days.

Because of the fixed orbit and low orbiting height of polar orbiting satellites, images can only be received from them by your ground station at certain times during the day and night, and these will only cover a small section of the earth's surface.

A *geostationary satellite* is one that is in a fixed orbit of 36 000 km above the earth and over the equator. At this height, every complete circuit the satellite makes lasts 24 hours, and so the satellite appears to stand still in the sky – or at least it would do if it could be seen!

Before satellites came on the scene, television pictures had to be broadcast from high towers. The *footprint* or transmission area of such towers was only about 30–40 km. Satellites have footprints too, though they are much larger and can cover the whole of a country. Meteosat, for example, can sense data from over a third of the globe, from its position over the equator above Africa. All the countries sensed by the satellite can receive images from it, since they are all within its footprint. Other geostationary satellites, however, such as those used for the satellite television channels, also carry spot beams which have relatively small footprints targeted over western Europe.

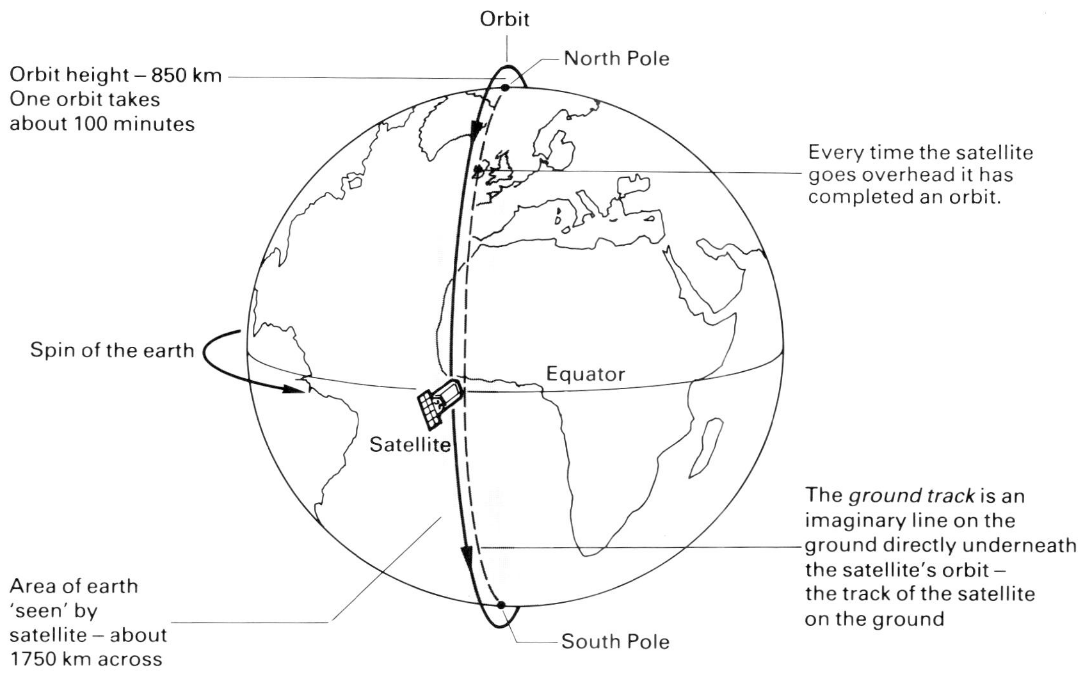

NOAA's orbit

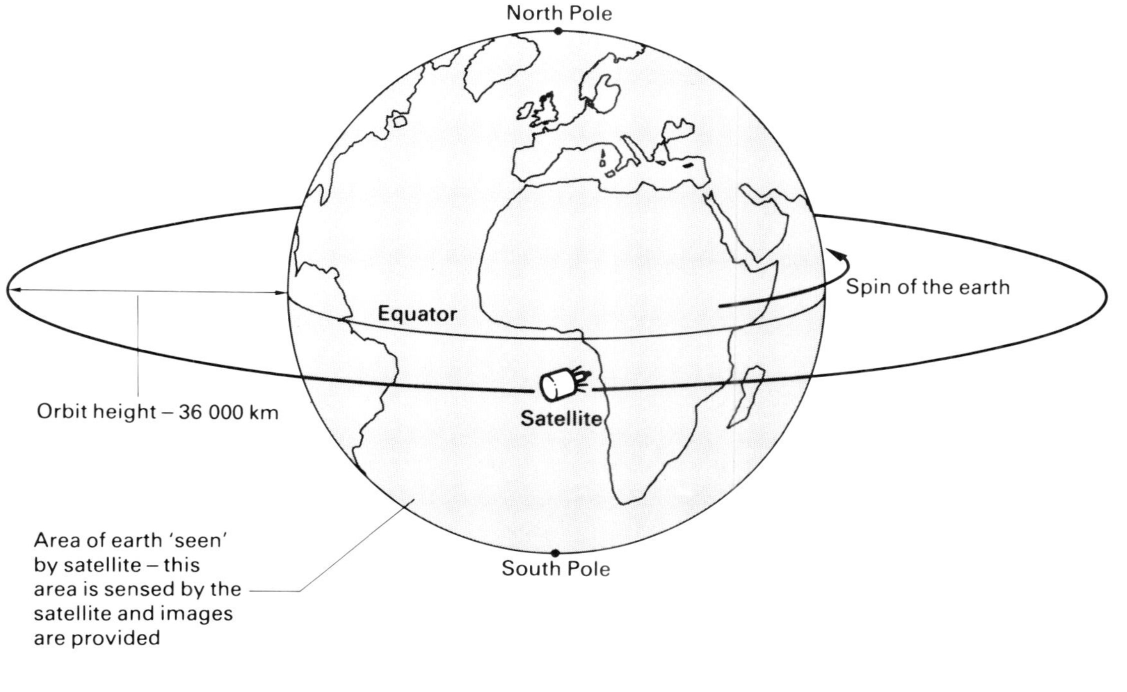

Meteosat's orbit

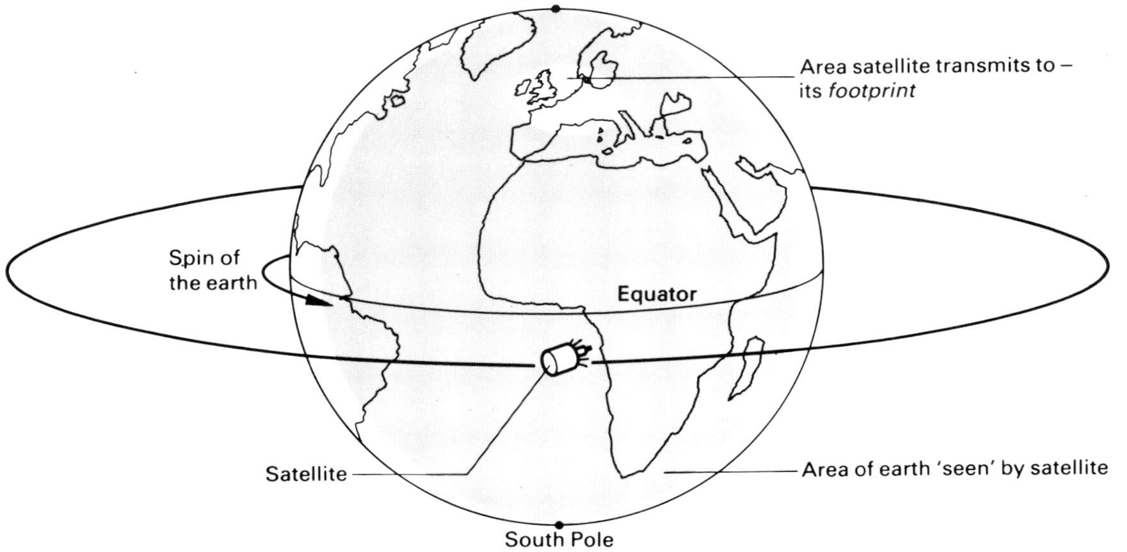

A geostationary satellite's footprint

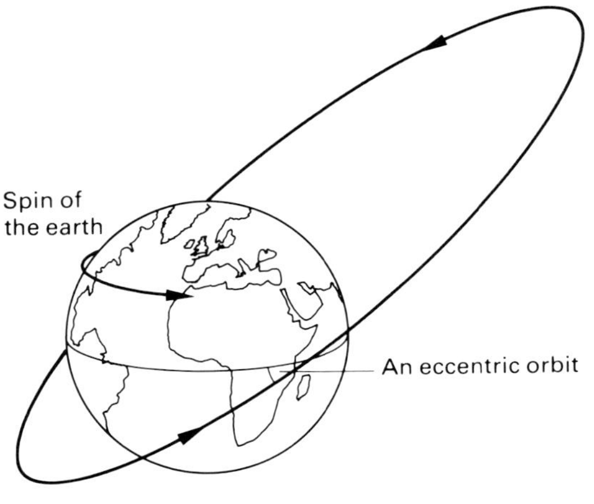

An eccentric orbit

For the study of the weather and for weather forecasting in the UK and Europe, Meteosat has many advantages over the NOAA polar orbiting satellites. To begin with, Meteosat can see a very much larger section of the earth, and this means that the whole of the North Atlantic is in view. In addition, Meteosat transmits images of Europe at regular intervals less than half an hour apart, and consequently the movement of weather systems can be tracked in detail, and future movement predicted with a fair degree of certainty.

The major advantage that NOAA has over Meteosat is that NOAA's image in the higher latitudes is nearly always overhead. For countries in such latitudes Meteosat is less useful because of the distortion of the image due to the curvature of the earth.

A third type of orbit is an *eccentric* one. Satellites of this type are often put up by the Soviet Union, because more of their ground track goes over the USSR on each pass of the satellite compared with the polar orbiters.

D Using Meteosat Images

Although Meteosat can 'see' over one-third of the earth's surface at a time, it does not transmit whole-earth images very often. Instead, it transmits data relating to smaller areas, which can be one of the two sizes shown on the next maps – the small C sections and the larger D sections. The sections are transmitted by the satellite in a set timetable during the day and night with the sections that are most relevant to European meteorologists, such as C2, C3, D1 and D2, being transmitted more frequently than the other sections. It is possible to purchase a printed timetable for Meteosat; alternatively, some pieces of software associated with commerical satellite receivers have the timetable. (See Appendix 7.)

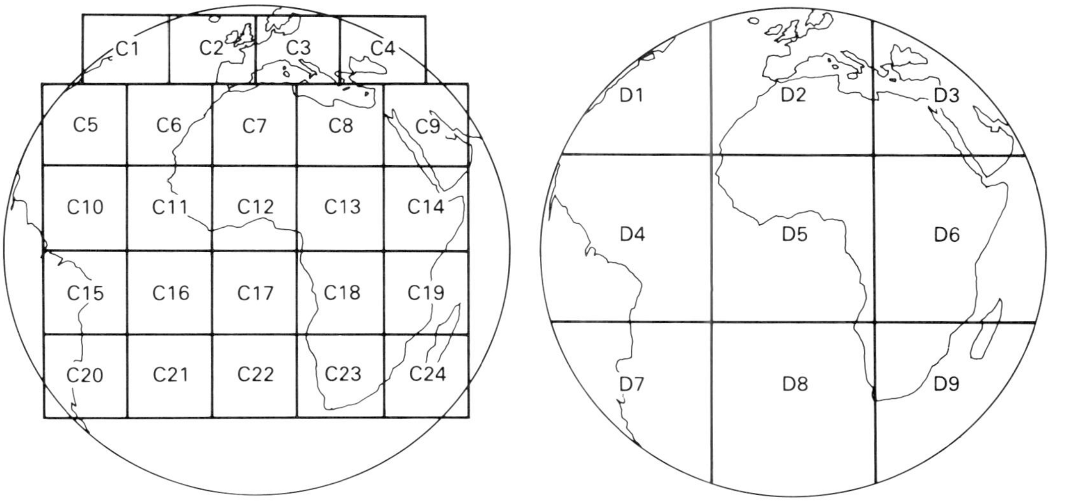

The sections used by Meteosat

The satellite has two channels. One is full for nearly 24 hours a day, transmitting an image every four minutes with only occasional breaks for administration and testing. The other channel is used less frequently, with only two or three images an hour being transmitted. It is on the latter channel, however, that the whole-earth images are transmitted during the day, and that images from GOES East – the geostationary satellite used by the USA – are beamed down, giving images of most of North and South America.

Meteosat's sensors detect both infra-red radiation being re-radiated and visible light being reflected from the earth's surface, and it sends down one image at a time for any one section of the globe. This contrasts with the NOAA polar orbiting satellites, which transmit both sets of data at the same time during a pass. Meteosat also sends down information on the water vapour content of the atmosphere.

The images from Meteosat are not what are called *real time* images – that is, the image being received is not 'live'. All Meteosat images are first transmitted to a ground station in Germany where they are cleaned up and the outlines of the countries are superimposed. The image is then sent back up to Meteosat, which rebroadcasts it for general use. As a result, each image is about 15 minutes 'late'. Again, this contrasts with images from NOAA, which are real time ones.

Reading a Meteosat timetable

It would be a good idea to set the students some activities using the Meteosat timetable whilst they are getting used to handling the satellite receiving equipment and being 'in charge' of the satellite ground station. The timetable below is explained overleaf.

GMT HH	**08**	**09**	**10**	**11**	**12**	**13**	**14**	**15**	**16**
MM	**CH1**	**CH1**	**CH1**	**CH1**	**CH1**	**CH1**	**CH1**	**CH1**	**CH1**
2			C3D	C9D	E3	C3D	C9D		C3D
6	D1		C4D	D1		C4D	D1		C4D
10	D2	D2	D2	D2	D2	D2	D2	D2	D2
14	CO2	CO2	CO2	CO2	CO2	CO2	CO2	CO2	CO2
18	CO3	CO3	CO3	CO3	CO3	CO3	CO3	CO3	CO3
22	TEST	D1	C5D	ADMN	D1	C5D	TEST	D1	C5D
26		D3	C6D		D3	C6D		D3	C6D
30		D4	D7		D4	D7		D4	D7
34	RANG	D5	D8	RANG	D5	D8	RANG	D5	D8
38	RANG	D6	D9	RANG	D6	D9	RANG	D6	D9
42	D2	D2	D2	D2	D2	D2	D2	D2	D2
46	CO2	CO2	CO2	CO2	CO2	CO2	CO2	CO2	CO2
50	CO3	CO3	CO3	CO3	CO3	CO3	CO3	CO3	CO3
54	TEST	C1D	C7D	E1	C1D	C7D		C1D	C7D
58	TEST	C2D	C8D	E2	C2D	C8D		C2D	C8D

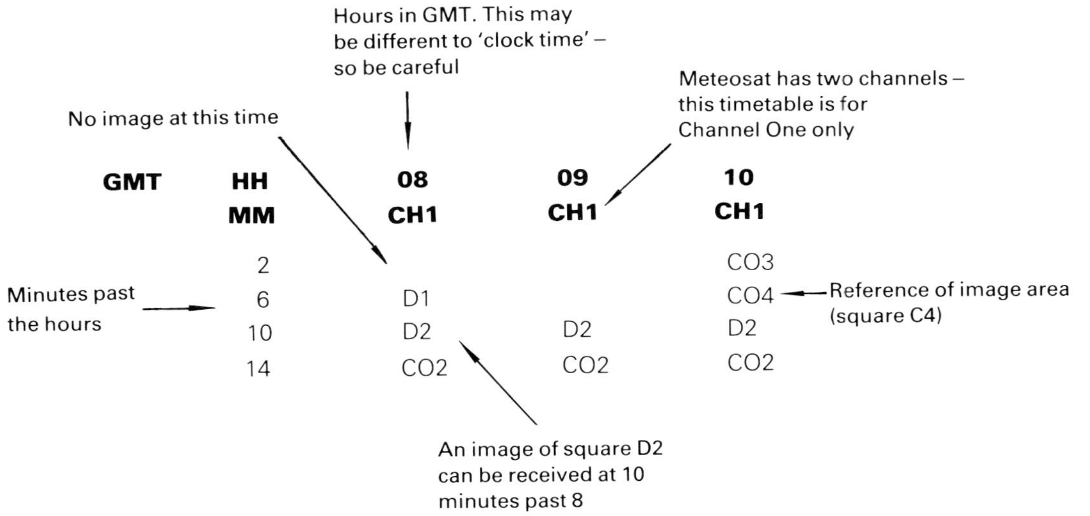

E Using NOAA Images

For a ground station located within the UK, the NOAA polar orbiting satellites can provide high-quality images of Europe and of large parts of the north Atlantic Ocean. The diagram below illustrates the possible coverage by the satellites during a 24-hour period, if the images taken on a number of passes are joined together.

One of the big disadvantages of NOAA compared with Meteosat is that only a limited number of images can be received each day. The fact that there is more than one NOAA satellite results in more frequent images than would be the case with a single satellite, but they never approach the number of the regularly spaced, frequent images from the geostationary satellite. Images from a polar orbiting satellite can only be received when the satellite's ground track passes through the *acquisition area* of the ground station – that is, when it is within range.

The time it takes for NOAA satellites to complete one orbit is around 100 minutes, and during this time the earth will have moved about 23° east. This means, therefore, that a ground station should receive about three or four good passes from each satellite every 24 hours.

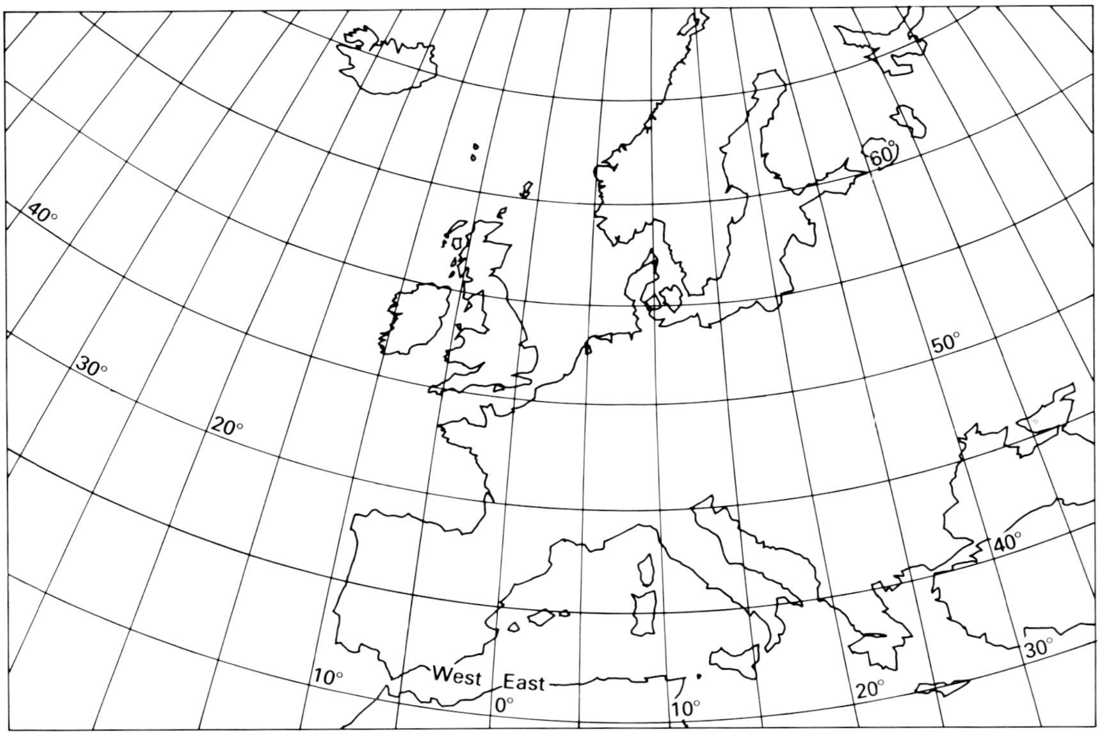

NOAA's possible coverage over 24 hours

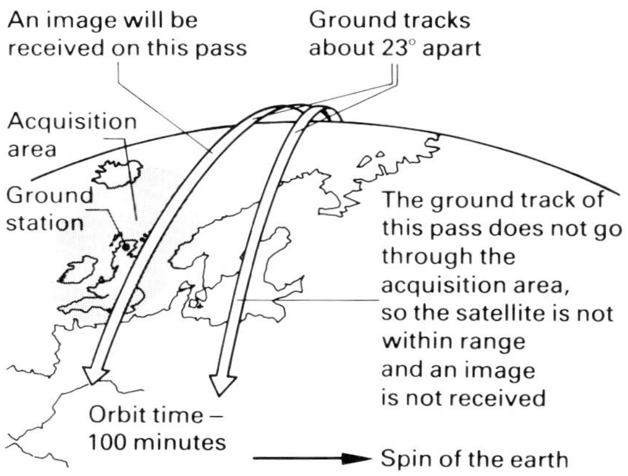

A polar orbiting satellite's reception area

The exact number of images a ground station will receive, however, will vary greatly, and depends on a number of factors. Perhaps the most important is the position of the satellite's ground track relative to the location of the ground station. The diagrams below show that in situation A, two images are received consecutively, 100 minutes apart, because both satellite passes are in range of the ground station. It must be said that this situation does not happen very often, though it is not rare. More usual, however, is situation B, where the ground track passes relatively close to

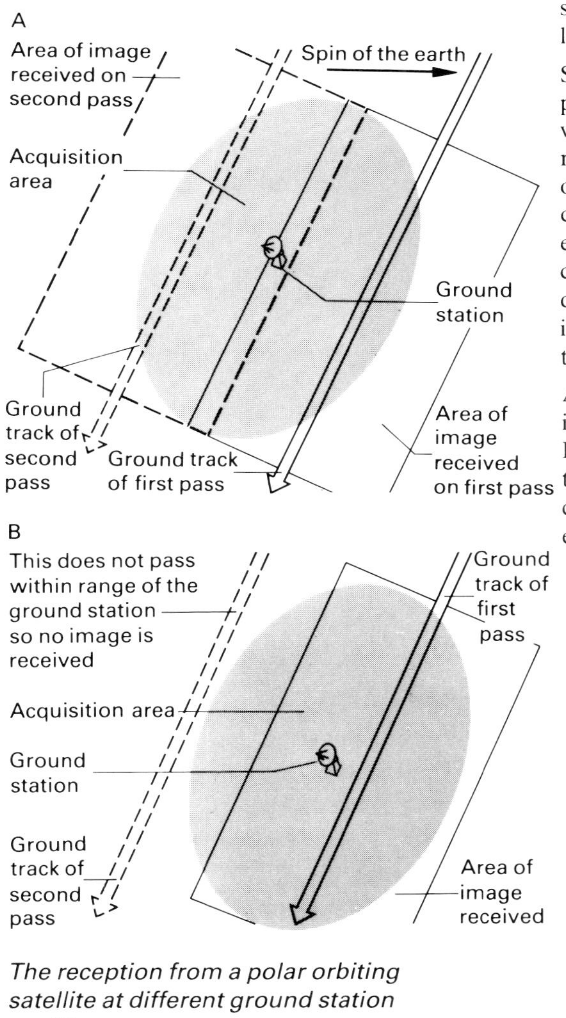

The reception from a polar orbiting satellite at different ground station positions

the ground station and only one image is received. The next would be in twelve hours' time.

Another characteristic of images produced by polar orbiting satellites such as NOAA is that each subsequent image will cover a different area of the globe. This is because consecutive ground tracks of each satellite pass will vary. One result of this fact is that it is not really possible to superimpose the outlines of countries on NOAA satellite images in the same way that is possible for Meteosat ones. Although some receiving systems do superimpose lines of latitude and longitude on NOAA images to help identify the area covered by the image, it is not always immediately apparent which part of Europe the image is of. When large areas of Europe are covered by cloud and little coastline is visible, interpretation of NOAA images becomes relatively difficult, especially for students.

An additional problem when using images from the NOAA satellites is that, because of the rotation of the earth, the satellites will sometimes pass overhead in a north–south direction, and sometimes in a south–north one. The diagram below illustrates why this happens. Some receiving systems compensate for this by making sure that all the images which appear on a monitor screen are the right way up. Other more basic systems may not do this, so the possibility is there of making some embarrassing mistakes concerning the location of the image!

Some satellite receiving systems will automatically pick up the images from the NOAA satellites, whereas others will need to be switched on to receive the image when the satellite is passing overhead. A number of orbital prediction computer software packages are available: these enable the pass times of the satellites to be calculated easily, and some even provide graphic displays of where the satellites are at any one point in time. Alternatively, a number of printed timetables are also available. (See Appendix 7.)

All in all, using images from the NOAA satellites is more difficult than using those from Meteosat. However, experience has shown that it is possible to use NOAA images with students in the classroom with just as high a degree of success and enthusiasm.

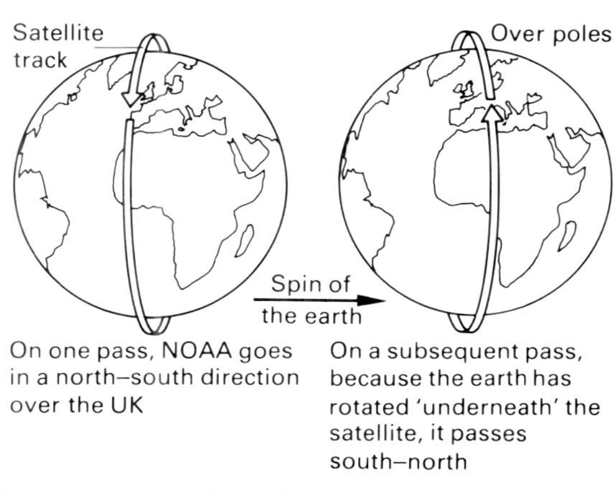

North–south and south–north passes of a polar orbiting satellite

F Notes on Sections 2–4 and Appendices 2, 5 and 8

Section 2

Nearly all the activities in this section are possible using even the most basic technology. The only problems might occur in the activities which involve the detailed identification of cloud types.

Some of the activities assume that both the infra-red and visible images for the same area at about the same time are available. In the case of NOAA, both types of image are transmitted at the same time, and on a number of receiving systems you can easily change from one to the other. With Meteosat, however, the visible and infra-red images are transmitted separately, and so one of the two images will have to be saved in some way – on computer disc, for example.

Section 3

When looking at depressions it is possible to use either visible or infra-red images. Often, it is useful to have both to look at.

The activities in this section deal with 'typical' depressions – if there are such things. Depressions are so variable, and some are more obvious than others. As a result, you might have to time your study of depressions carefully to link in with a good set of satellite images. On the other hand, one of the objectives of the activities in this pack is to give the students the skills to interpret any satellite image with reasonable success, so do not just restrict their study to the 'good' images of depressions.

Depressions can be of different strengths, depending on the pressure. The 'best' depressions will pass over the UK during the autumn, winter and spring.

Appendices 8 and 9 are particularly useful for this section.

Section 4

All the activities in this section assume that you are using infra-red images, as these images give an indication of temperature.

The images used have to be of high quality and high resolution. Some of the more basic satellite receiving systems will not produce worthwhile images for these activities.

The images used will have to be selected carefully. Not every infra-red image will show something worthwhile in terms of these activities. It might be useful, therefore, to build up a library of suitable images – if you are able to store your images in some way.

The best images to use are those from high pressure situations, in which large areas of land can be seen. In these conditions the infra-red sensors will pick up variations in ground temperature.

Appendix 2

Base maps have been provided for NOAA, Meteosat and GOES East images.

The NOAA base map is a general one, and the first part of any activity which involves the transferring of information from an image to the base map should be to mark on the map clearly the area of the North Atlantic and Europe actually 'seen' on the image. The intersections are 5° apart.

The Meteosat base maps have been provided for most of the sections for which Meteosat transmits images (D1–9 and C2–3 – see the diagram on p. 7).

You should note that there are problems when using most of the maps, in terms of scale and distance. Obviously, the scale of the map will change both within and between maps, especially on the higher latitude Meteosat images. This problem should be pointed out to the students, who should be encouraged to make relatively generalised distance measurements. Accuracy to the last kilometre is not possible.

To help with location, one latitude and longitude intersection is numbered. The crosses marking the intersections are 10° apart on all the Meteosat maps. The scale marked on the maps refers to the approximate distance from the numbered intersection to the next 10° intersection west or east. It should be adjusted slightly if distances to the north or south are being measured.

Appendix 5

The weather map for Images 3 and 4 (1 August 1979) is given in 2.7 Activity 1.

The maps were produced in a different format after 1979, and do not cover the same area as the earlier ones.

Appendix 8

As an extra to this pack, it is possible to purchase from the publishers large plastic wall maps showing the UK and western Europe. These maps can be used together with the symbols in this appendix to display weather forecasts and the like, and provide a useful additional resource for a range of classroom activities. Alternatively, the symbols could be used with school-produced weather display maps.

1.1

What Do Satellites Do?

A satellite is simply an object which is spinning in an orbit around a planet or a star. The moon is an example of a natural satellite spinning round the earth. However, when you think of satellites, you will probably think first of those made by people, rather than of things like the moon. The first satellite made by people was launched in 1957, and there are now thousands of them in space used for all sorts of purposes.

Geographers use the information from a large number of different satellites, and although it is probably the weather satellites that are the best known, they provide information on many other subjects as well.

Spy satellite reads *Pravda*!

A new spy satellite launched by the USA has cameras that are so powerful that it can take pictures of people! It is said that a picture has been taken of a man in Moscow sitting on a park bench reading a copy of *Pravda*.

With the Star Wars programme the USA is also building 'killer' satellites which can destroy enemy satellites in space.

Flood control by satellite

When snow melts, the water can cause flooding in rivers. Now a possible answer is on the way as images from satellites help scientists work out snow depths, and therefore the likelihood of flooding.

Satellite TV for all in the 1990s

Satellite television is about to take off. By the mid 1990s most households will have a small dish aerial on their roof to receive television pictures from any one of hundreds of television channels.

Messages around the world

Business people now use satellites to send messages around the world at high speed. Copies of documents and pictures can also be beamed across the globe.

A hunter in space catches fish!

Fishing has become much easier during the last few years, as a new generation of fishermen use satellite technology!

Satellites which can map the temperature of the sea can help predict where shoals of fish are to be found.

Pollution up in the Gulf

As a result of the Gulf War between Iran and Iraq, the level of oil pollution in the Gulf is on the increase. Recent satellite images have picked up a huge oil slick moving slowly down the Gulf.

New oilfield discovered

A set of satellite images has led to the discovery of an oilfield.

Wonder technology!

Did you know that satellites can handle literally thousands of telephone conversations at the same time?

Crop survey gives hope for Africa

Recent satellite surveys of crops and crop yields in parts of Africa are giving hope that a recovery may be on the way. Satellite pictures taken show that crop yields are likely to be up on last year.

Sailing is safer

A new network of navigation satellites has made the seas safer. Ships can plot their position more accurately using these satellites than ever before. They can also be used for ships to contact the rescue services in an emergency.

1.1

ACTIVITIES

1. Read each of the newspaper extracts above, and for each one say how satellites have been used to help people.
2. Which of the satellites in the newspaper extracts do you think is the most useful to people? Why did you choose that one rather than the others?
3. If you could design a new satellite which would be useful to people, what would it do?
4. Try to find out the names of some of the satellites in space.
5. Find out how your life is influenced directly and indirectly by satellites. For example, have you watched a television programme recently that has been transmitted by satellite from another country?

OBJECTIVES

CHECKLIST

1.1

What Do Satellites Do?

Do you

Tick here

- know about the different uses of satellites? ☐

Can you

- use a library to find out things about satellites? ☐
- choose the most important use of satellites, in your opinion? ☐
- suggest a new use for satellites which would be helpful to people? ☐

2.1

What the Images Mean – Visible

Many weather satellite receivers will give you two different kinds of image of the same area. It is important that you realise what kind of image you are looking at because each one will tell you different things about the weather.

A *visible image* is a measure by the satellite of how much light is reflected by the earth as the satellite passes overhead. Clouds reduce the amount of sunlight reaching the earth's surface and appear relatively bright on a satellite visible image, because they reflect a lot of light back to the satellite.

Thick clouds will reflect a lot of sunlight whilst thin clouds, because they will let some sunlight through, will appear a dull white or grey on the satellite image. In other words, the brighter the white on the image, the thicker the cloud.

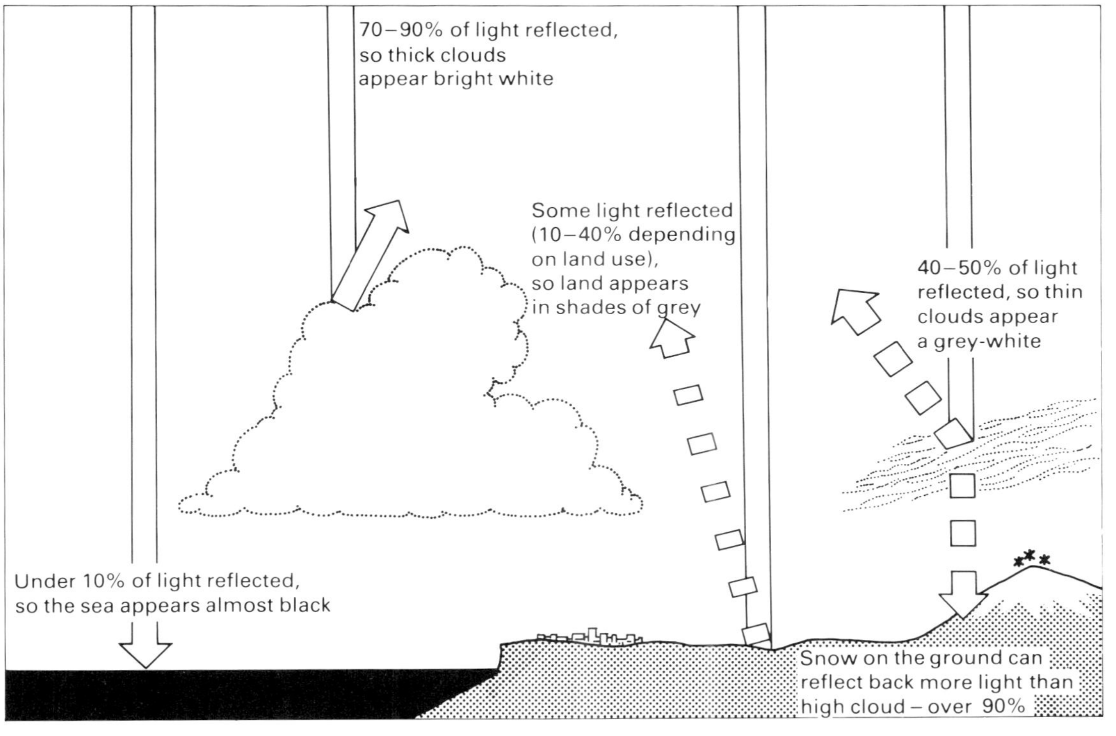

Both land and sea, on the other hand, don't reflect much light from the sun, and therefore appear in shades of grey on the satellite image, with the sea usually appearing darker than the land.

There are a number of points you should remember when looking at visible images:

1 You obviously won't be able to see visible images during the night-time!

2 In the early morning and late afternoon, only parts of a visible image may appear – depending, of course, on the area covered by the image. This is because in some parts of the area covered by the satellite it is still night-time whilst in others it is daylight.

3 If the image is for either the early morning or late afternoon, then there may not be enough light reflected by the earth to give a bright image, even if it appears bright on earth.

2.1

The sketch below is of the thickness of cloud on Image 1. Look at the two alongside each other – they should help you to interpret visible images.

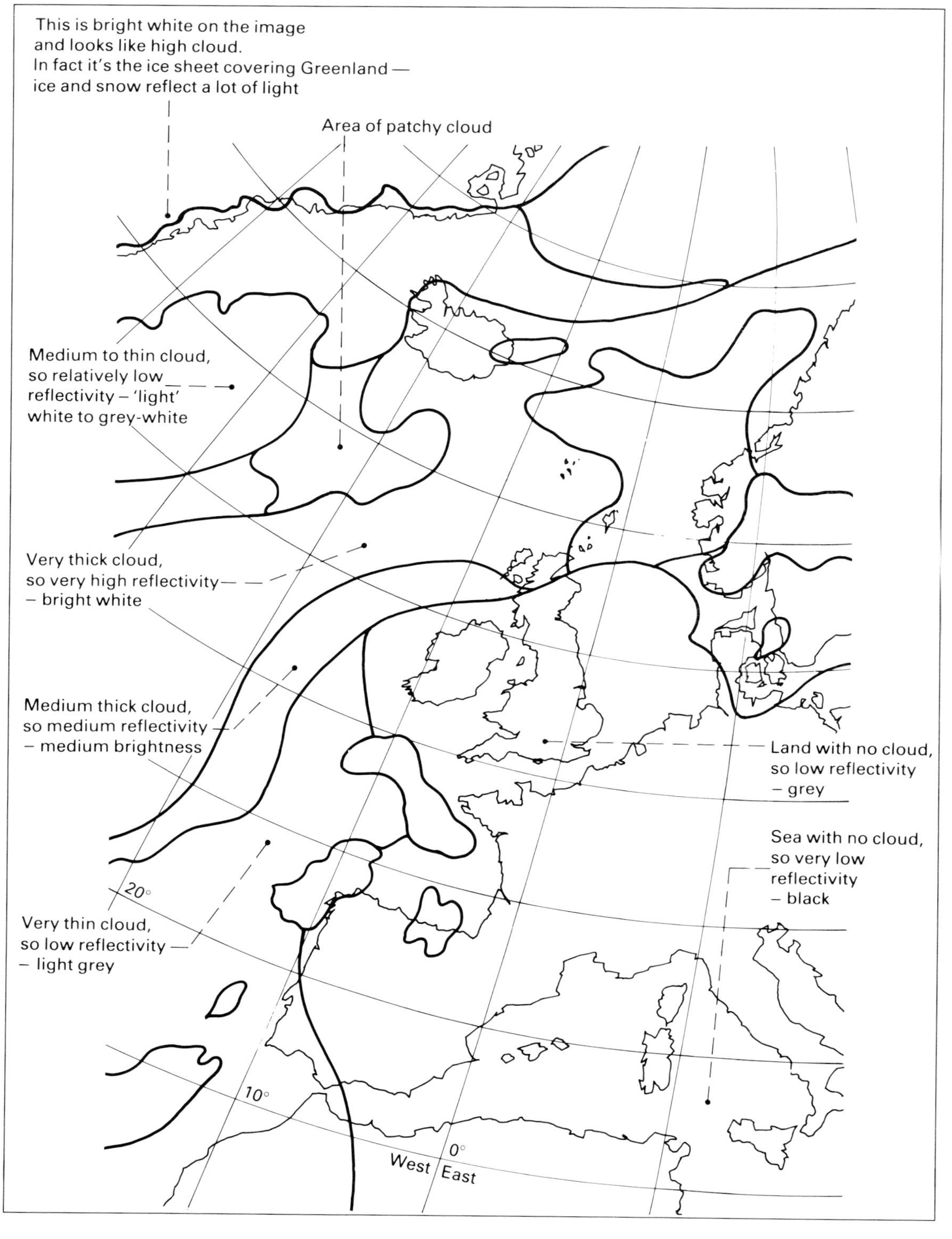

Cloud thickness on Image 1

2.1

ACTIVITIES

1 Look carefully at the types of cloud on the diagram below. The most important cloud types are shown. For each type estimate the following figures, using the diagram. (Be careful since the scale is not linear.)

a the height of the base of the cloud;
b the height of the top of cloud; and
c the thickness of the cloud.

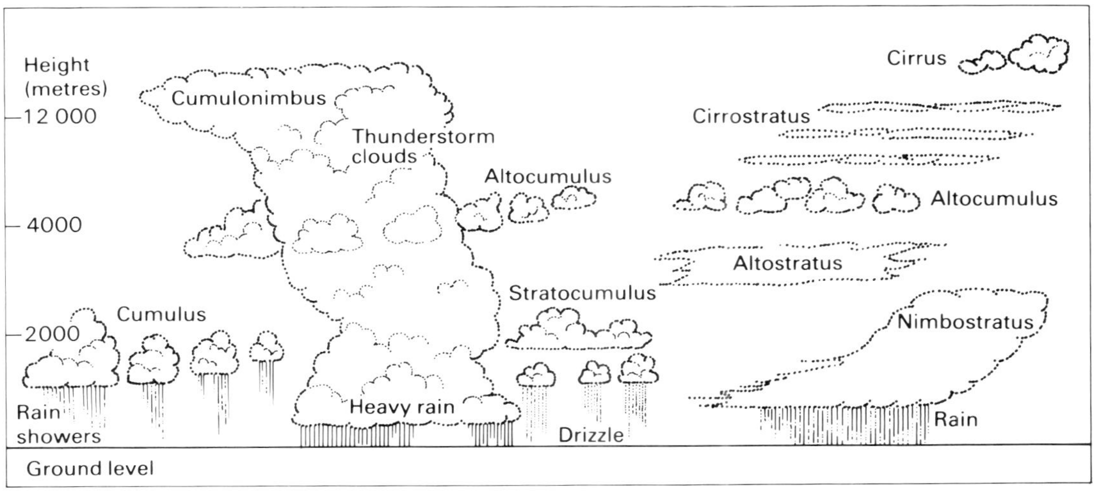

2 Look at either Image 3 or the current live satellite image. On a copy of the right base map, draw a line around the area seen by the satellite and then find the areas of thickest cloud cover.

Shade these areas on your map and then name any five countries that are on the image. Work out whether these countries have a thick cloud cover or not.

3 You could also try working out where the areas of scattered cloud are on your image, and which areas have little or no cloud at all.

2.2 What the Images Mean – Infra-red

A second kind of image you will be able to see is an *infra-red image*. The diagram below shows the electromagnetic spectrum, and both NOAA and Meteosat can sense waves of different lengths within this spectrum. They can pick up not only visible light (the sort we would be able to see if we were in orbit around the earth!), but also the amount of *infra-red radiation* re-radiated by things on the earth. This radiation has a different wavelength – one we can't see, but the satellites can sense.

Everything on the earth's surface gives off infra-red radiation and the sensors on the satellite can measure this. Warmer surfaces tend to give off more radiation than cooler ones, and so infra-red images are indirectly a measure of temperature.

On the satellite image, the colder a surface is, the whiter it will appear. Warmer surfaces will appear in shades of grey.

As temperature normally gets lower the higher you go, clouds will usually appear white on satellite images, whereas the land and the sea, being warmer, will appear as darker shades. Clouds appear white because they give off little radiation – the brighter the white, the higher the clouds.

The coastline will often stand out clearly on an infra-red image because of the different temperatures of the land and sea.

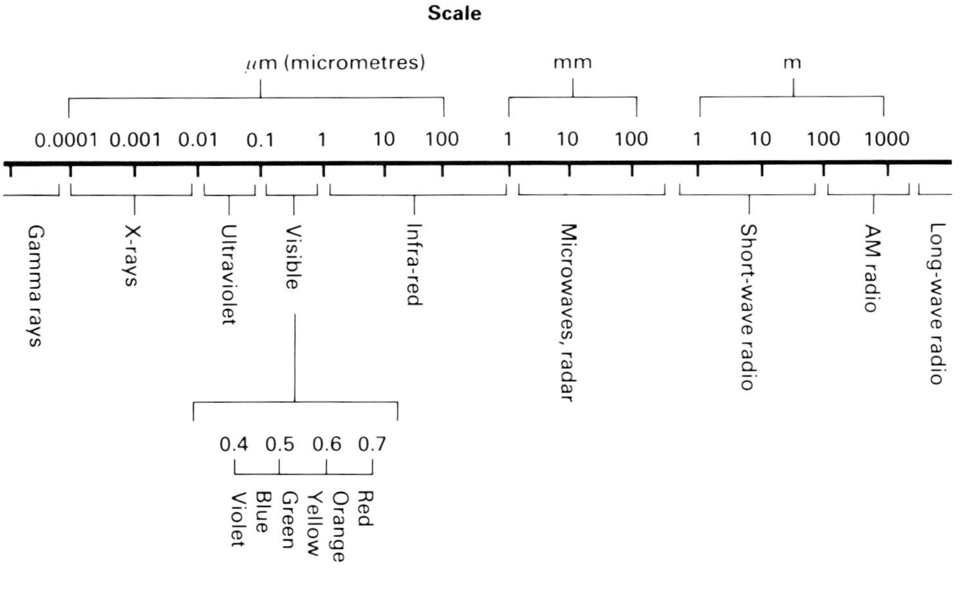

The electromagnetic spectrum

It is important to realise, however, that the shades of land and sea will change as their temperatures change. In the winter, it is often the case that the sea is warmer than the land, especially in the early morning. The sea will be a darker shade on the infra-red image. In the summer, the land is nearly always warmer than the sea, and so the land appears in a darker shade. At some times of year, both land and sea are about the

2.2

same temperature, so they will both be the same shade on the infra-red image. If your receiving system doesn't put some kind of grid lines on your image, then it will be difficult for you to tell where places are on images like this!

As well as changes through the year, there are also often changes in shade between the day and night as land and sea heat up and cool down.

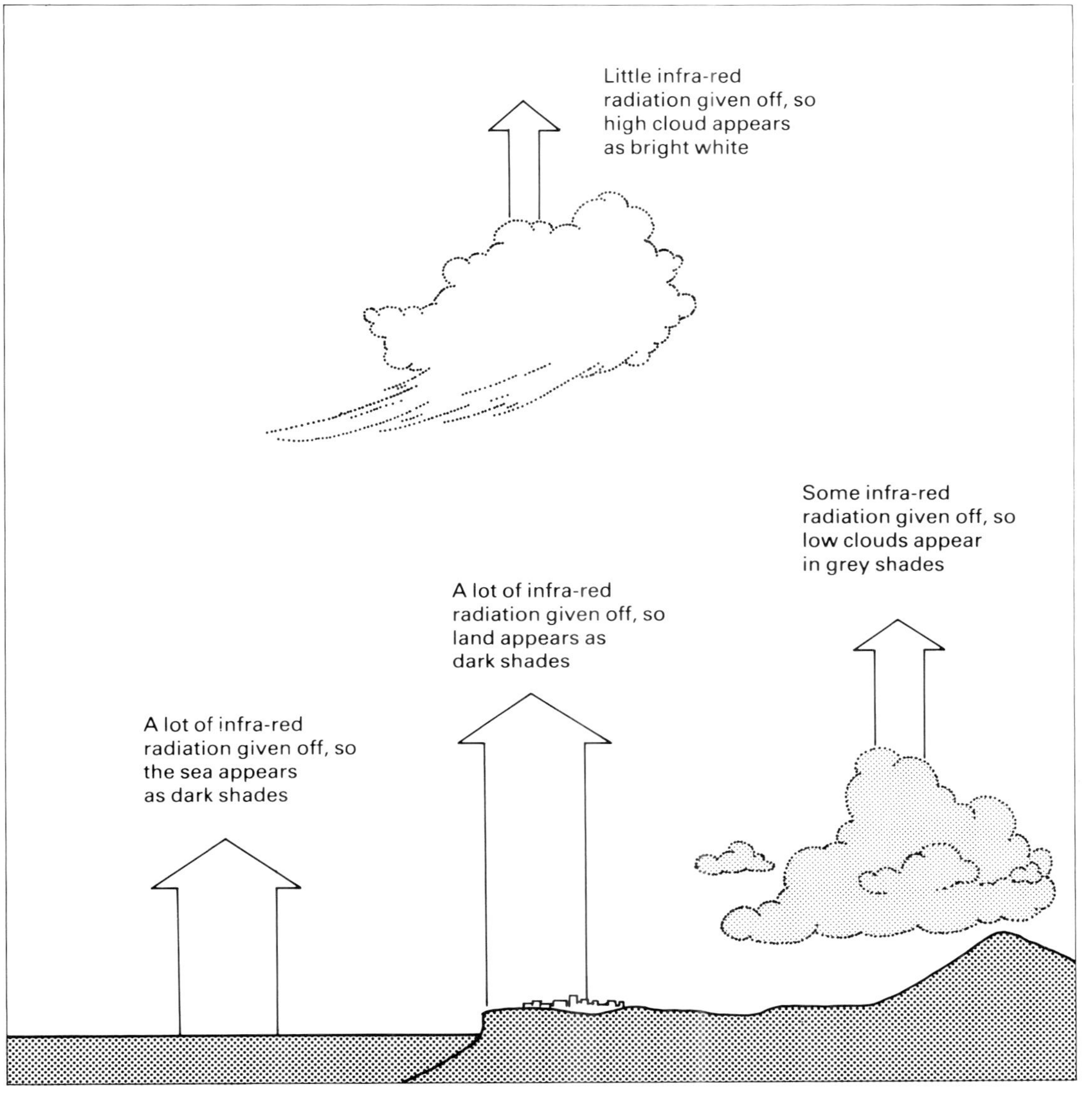

There are a number of points to remember when looking at infra-red images:

1 Infra-red images can be received at night-time.

2 An infra-red image doesn't tell you a lot about the thickness of clouds. Thin clouds that are high up will look much the same as the tops of very thick clouds which happen to be at the same height.

The sketch on page 18 is of Image 2. Look at the two alongside each other – they should help you to interpret infra-red images.

2.2

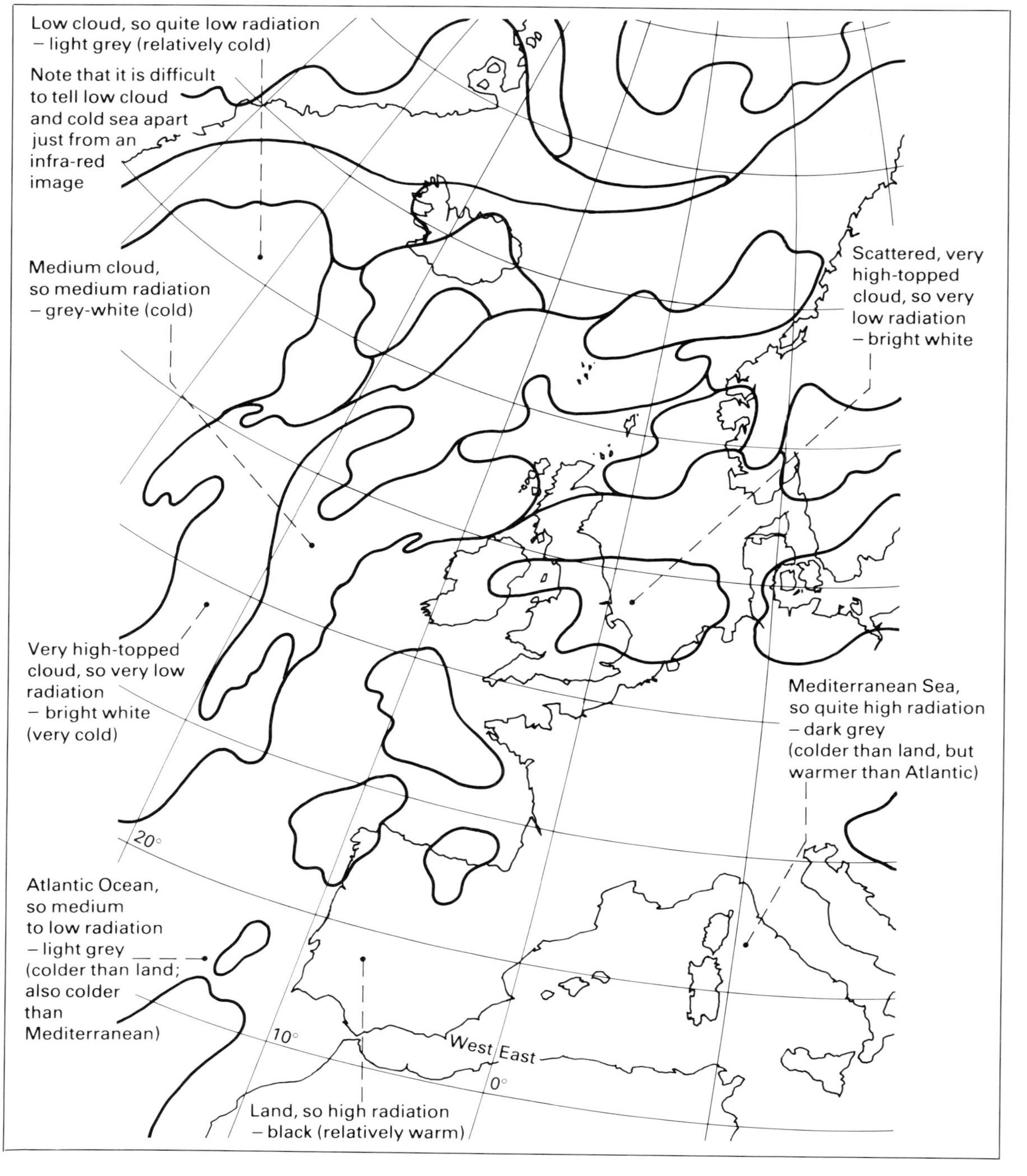

Cloud height on Image 2

ACTIVITIES

1 Look carefully at the graph on page 19. It shows how the average temperature of the atmosphere will change with height from the ground.

For each of the cloud types you looked at in 2.1 Activity 1, use the graph on page 19 to work out the temperature of the base and tops of the clouds. You should assume that the temperature of the ground is 18 °C.

2.2

How temperature declines with height

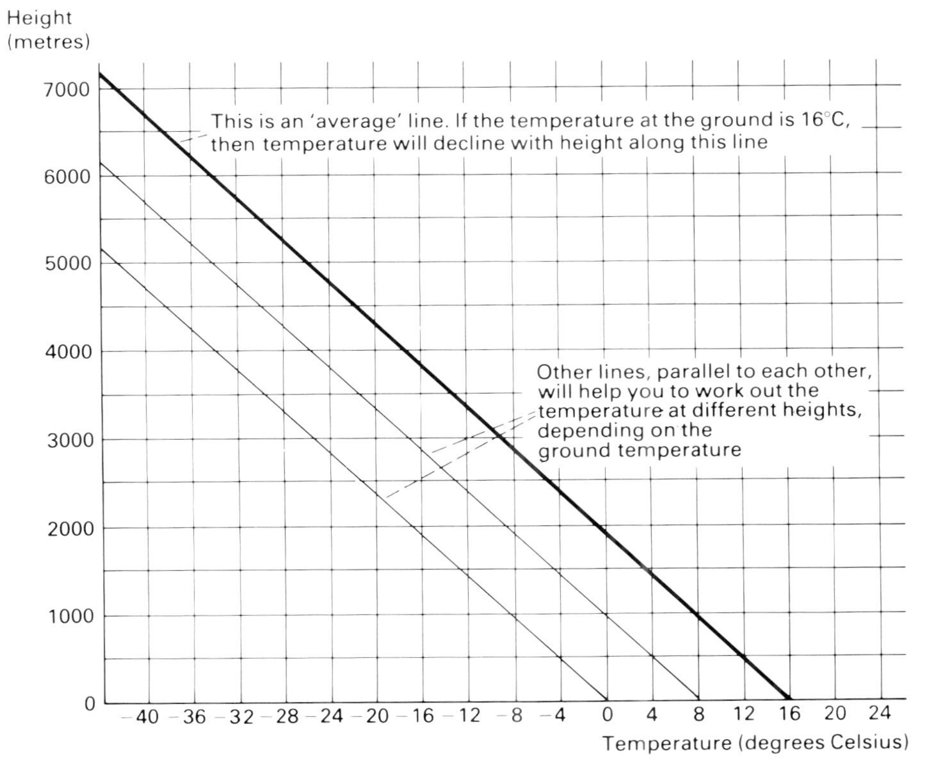

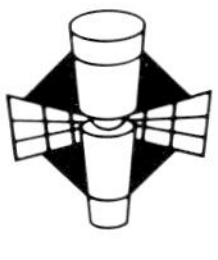

2 Look at either Image 4 or the current live infra-red image. On a copy of the right base map, draw a line around the area seen by the satellite. Find out where the areas of high-topped, medium-topped and low-topped clouds are on the image and mark these on your map.

If possible, use an atlas to help you name a country which is covered by each height of cloud.

2.3

How are Clouds Formed? (1)

The air around you contains water. Usually, it is in the form of an invisible gas called water vapour. Warm air can hold quite a lot of water vapour, whilst cold air cannot hold as much. The amount of water vapour in the air is known as the *humidity*.

If warm air is cooled, it will reach a temperature where it can't hold all the water vapour it has in it. When it gets to this temperature, the water vapour changes into tiny droplets of water. This process is *condensation* and the water vapour is said to *condense*. The temperature at which this happens is called the *dew point temperature*.

When there are lots of small water droplets close together, a cloud will be formed. It will rain when the drops get too heavy to stay in the cloud.

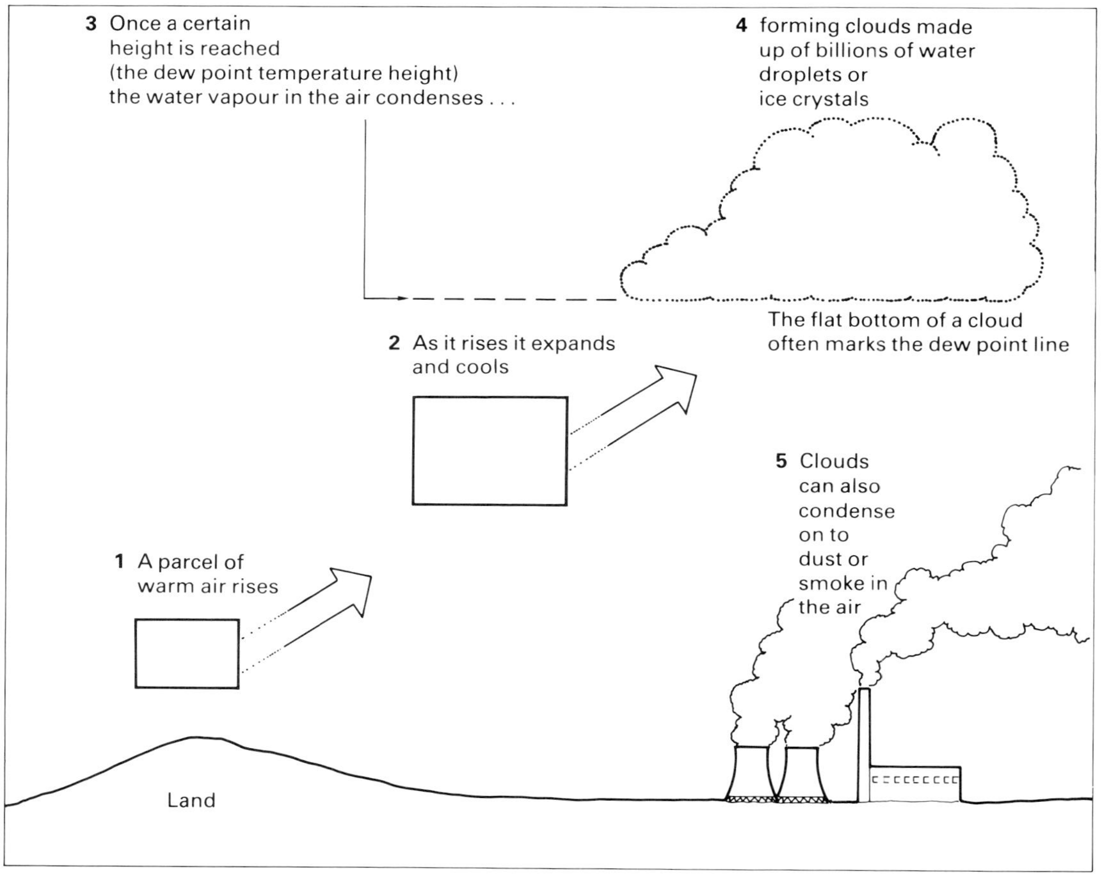

There are basically two different kinds of cloud. 'Heaped' clouds are called *cumulus* clouds. They form when bubbles of air rise quite quickly and cool. Wispy or layered clouds are called *stratus* clouds. These tend to form when a larger mass of air rises slowly.

Clouds are also found at different heights, and sometimes the name of the cloud tells you this – the word *alto*, for example, tells you the cloud is at a medium height. Some clouds bring rain more than others. The word *nimbus* in front of the cloud name or after it tells you it is a rain cloud.

The diagram for 2.1 Activity 1 shows you different clouds.

2.3

ACTIVITIES

1 Look at the sketches of clouds below. What do the descriptions 'altocumulus' and 'cumulonimbus' tell you about the clouds?

2 Make a copy of the sketch, and then work out what each cloud type will look like on a visible (V) image and on an infra-red (IR) image. Fill in the boxes beside each cloud type with your description.

3 Look at the sketch on page 22 together with Images 1 and 2. It should help you to use the information you have put on your diagram to identify actual cloud types on satellite images.

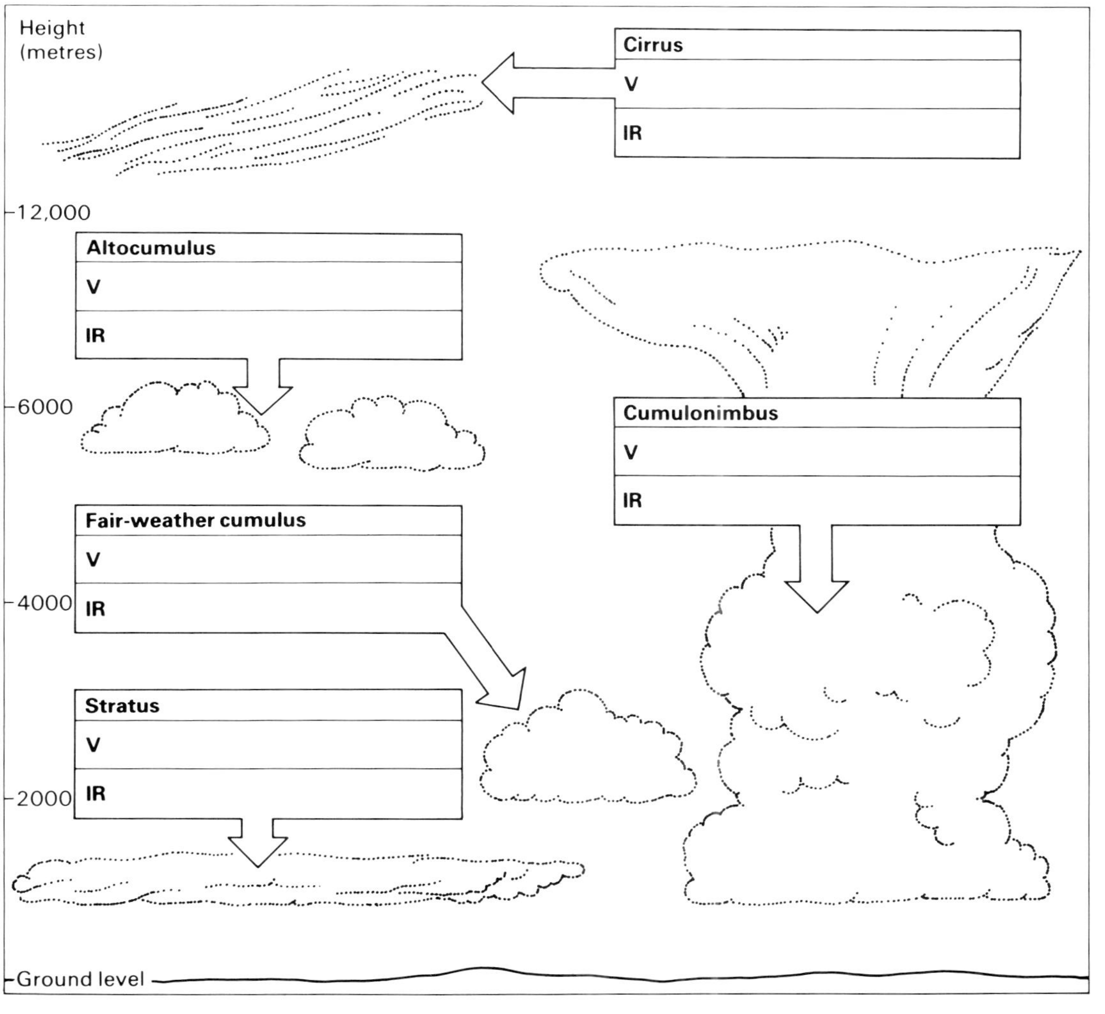

4 How do you feel about different kinds of cloud and the weather that comes with each type? Write a few lines about the different clouds you see, to show how the weather affects your moods.

5 Try to find out why clouds are different shapes.

2.3

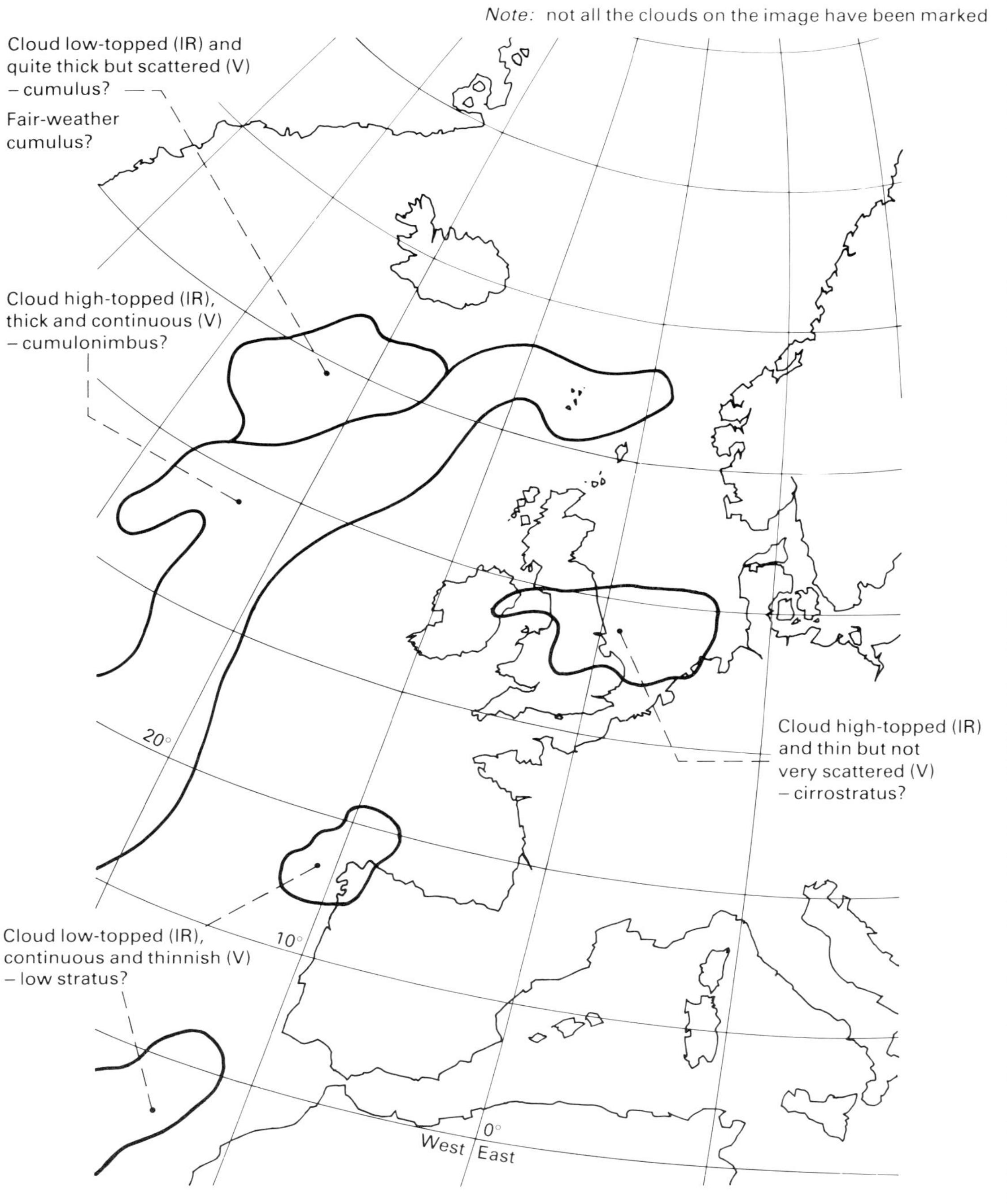

Types of cloud on Images 1 and 2

2.4 How are Clouds Formed? (2)

You have already seen that clouds form when air containing water vapour is cooled enough. This cooling happens in three ways, and these are shown in the diagrams below.

A When the ground is heated, the air above the ground is warmed and begins to rise. As it rises, it cools and clouds may form – this is called convection. If these clouds bring rain, it is called *convectional rainfall.*

B When air hits the side of hills or mountains, it is often forced to rise up them. As the air rises, it is cooled and clouds form. If these clouds bring rain, it is called *relief rainfall* or *orographic rainfall*, because it has to do with the height of the land.

C Warm air can also be cooled when it comes into contact with cooler air itself. This happens along a front line, and if these clouds bring rain it is called *frontal rainfall.*

A

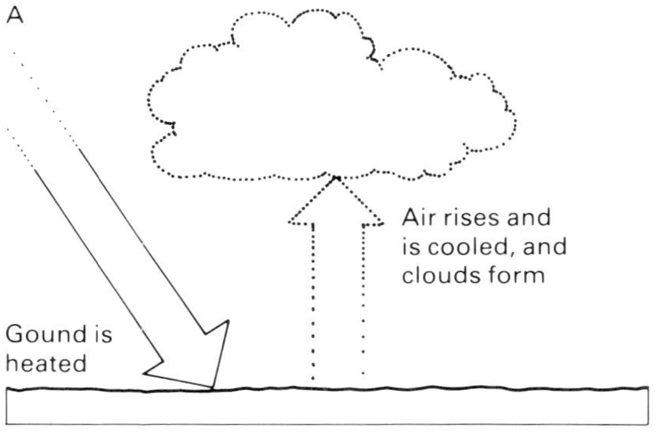

Look at Image 1 – you can see examples of this kind of cloud over Italy on this hot summer's day.

B

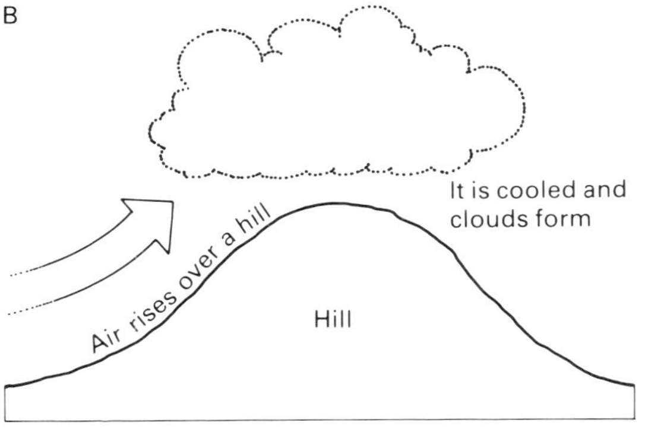

Look at Image 6 – you can see clouds like this over many of the upland areas of the UK.

C

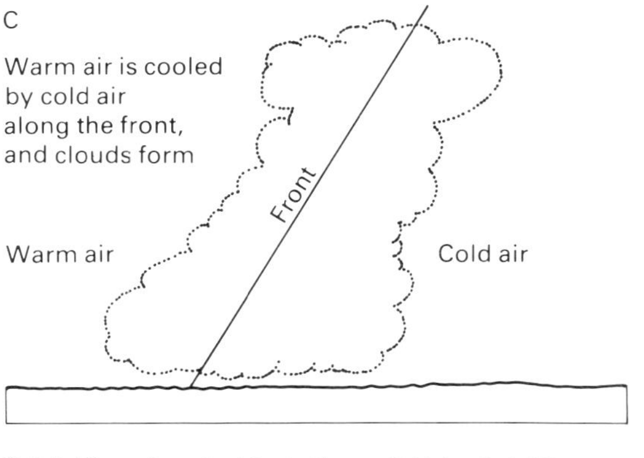

Look at Image 5 – the long line of cloud is a front between air with two different temperatures.

2.4

ACTIVITIES

1 Do this activity either using Images 3 and 4 or using the current live visible and infra-red images on your own receiving system.

Look at the visible image, and mark on a copy of the right base map any areas of thick clouds – those that appear bright white on the satellite image. Shade these areas in a light colour on your map.

Now look at the infra-red image, and mark on the same map, but using a different colour, any areas of high clouds – those that appear bright white on the satellite image.

Where the two areas of shading overlap there are probably thick cumulus clouds, which are associated with rain.

If you are using live images, you can check to see whether you identified the clouds correctly if it happens to be a day when the cumulus clouds are over your satellite receiving station.

2 Do this activity either using Images 3 and 4 or using the current live visible and infra-red images on your own receiving system.

Look at your results for 2.3 Activity 3, and then fill in a copy of the checklist below as you are able to find the different kinds of cloud.

Cloud type	**Found (✓)**	**Date found**
Cirrus		
Fair-weather cumulus		
Cumulonimbus		
Stratus		
Altocumulus		

3 When reading novels, or other books, make a note of any scenes in which the author describes the weather. Can you work out what kind of cloud the authors are describing?

A good book to start with would be *Wyatt's Hurricane* by Desmond Bagley. As the title suggests, this is about a hurricane that devastates a small island.

2.5

Clouds for Real

Different kinds of cloud tend to be associated with different kinds of weather. How good are you at linking particular patterns on a satellite image with different kinds of weather?

ACTIVITIES

1 Don't look out the window before doing this activity.

Look at a satellite image soon after it has been received and say what kinds of cloud should be in the sky at your location. Then look out of the window to see if you were correct. Try not to cheat!

2 If the temptation to look out of the window is too much for you, then try this activity.

On a copy of the right base map, make a detailed weather map for the UK showing the kinds of cloud. Also mark on your map where one or two of your relatives live.

Either in school, or when you get home, phone up your relatives, and tell them what kinds of cloud they had over them during the day! It might be a good idea to warn them the night before that you are going to do this, so that they will make a special effort to look at the weather.

Did you get the right answers?

3 Look at the clouds outside today. Imagine you are talking to a friend over the phone, and describe briefly what the clouds and weather are like.

4 Using both infra-red and visible satellite images for each day, fill in a diary like the one below. This will help you to understand the relationship between satellite images and what is happening in real life.

Day of the week	**Weather**	**Infra-red image**	**Visible image**
Monday			
Tuesday			

2.5

5 Look at Image 3. Mark on a copy of the right base map the parts of Europe you think are going to be cloudy.

Now look at the table below, which shows information from European weather stations for the same day on how much cloud was covering the sky at each place. The figures refer to how many eighths of the sky were covered with cloud – 8 means total cover, and 0 means clear skies. Compare your cloud map with the actual weather data. Did you put the clouds in the right places?

Gibraltar	0	Lerwick	8
Lisbon	5	Stornoway	7
Madrid	2	Belfast	7
Valencia	2	Dublin	7
Barcelona	1	Oslo	6
Bordeaux	4	Copenhagen	5
Paris	8	Cologne	6
Channel Islands	8	Munich	7
Scilly Isles	8	Vienna	7
London	5	Warsaw	6
Nice	1	Belgrade	0
Genoa	8	Dubrovnik	0
Rome	5	Athens	0
Algiers	0	Istanbul	0
Venice	1	Bucharest	0
Geneva	1	Kiev	8
Amsterdam	8	Moscow	8
Aberdovey	3	Stockholm	8
Manchester	7	Helsinki	7
Ayr	7	Prague	7
Aberdeen	7		

COURSEWORK ACTIVITY

If you are able to make a hard copy (print-out) of your satellite images, you could make a collection relating to different types of weather.

You could then design a poster which other students could use to help them interpret what the satellite images mean.

It would be a good idea to put photographs or sketches of the clouds on the poster as well.

2.6 What about Rainfall?

It isn't possible just from looking at a satellite image to say definitely that it is raining at a particular location, or that rain is on its way. As you know, forecasters get things wrong!

However, it is true to say that certain kinds of cloud are more likely to bring rain than others. For example, thick clouds usually mean rain. Sometimes these are low clouds such as nimbostratus clouds. Sometimes the clouds have low bases but extend a long way up, such as cumulonimbus clouds.

Other kinds of cloud, such as cirrus or fair-weather cumulus clouds, don't as a rule bring rain with them.

You should already have had some practice at recognising these kinds of cloud.

ACTIVITIES

1. Look at the current live visible and infra-red images of the UK and work out whereabouts you would expect it to be raining.

   You can test your interpretation of the satellite images as you did in 2.5 Activity 2.

   You could also test your results by comparing them with the weather maps provided during the day by television weather forecasts. Alternatively, your results could be compared with the weather maps on Ceefax and Prestel, if your school has this facility.

2. You could also try drawing rainfall maps for Europe. You will have to use other sources of weather information to check on the accuracy of a European weather map.

   The Times, *The Guardian*, the *Independent* and the *Daily Telegraph* are all newspapers that print lists of rainfall for selected places in Europe. Although the actual rainfall figures aren't printed, the weather report usually gives a description such as 'rain' or 'cloudy'. You can use information of this kind to draw a generalised weather map, such as the one on page 28.

   Look at Images 3 and 4, and on the right base map draw a simple rainfall map for Europe. Now look at the rainfall information below, which shows where it was raining in Europe on the day the image was taken. Make a rainfall map similar to the one on page 28, and compare both maps you have drawn. Did you put the rainfall in the right places?

Scilly Isles	London	Brussels
Channel Islands	Prague	Oslo
Copenhagen	Paris	Manchester
Berlin	Amsterdam	Stornoway

2.6

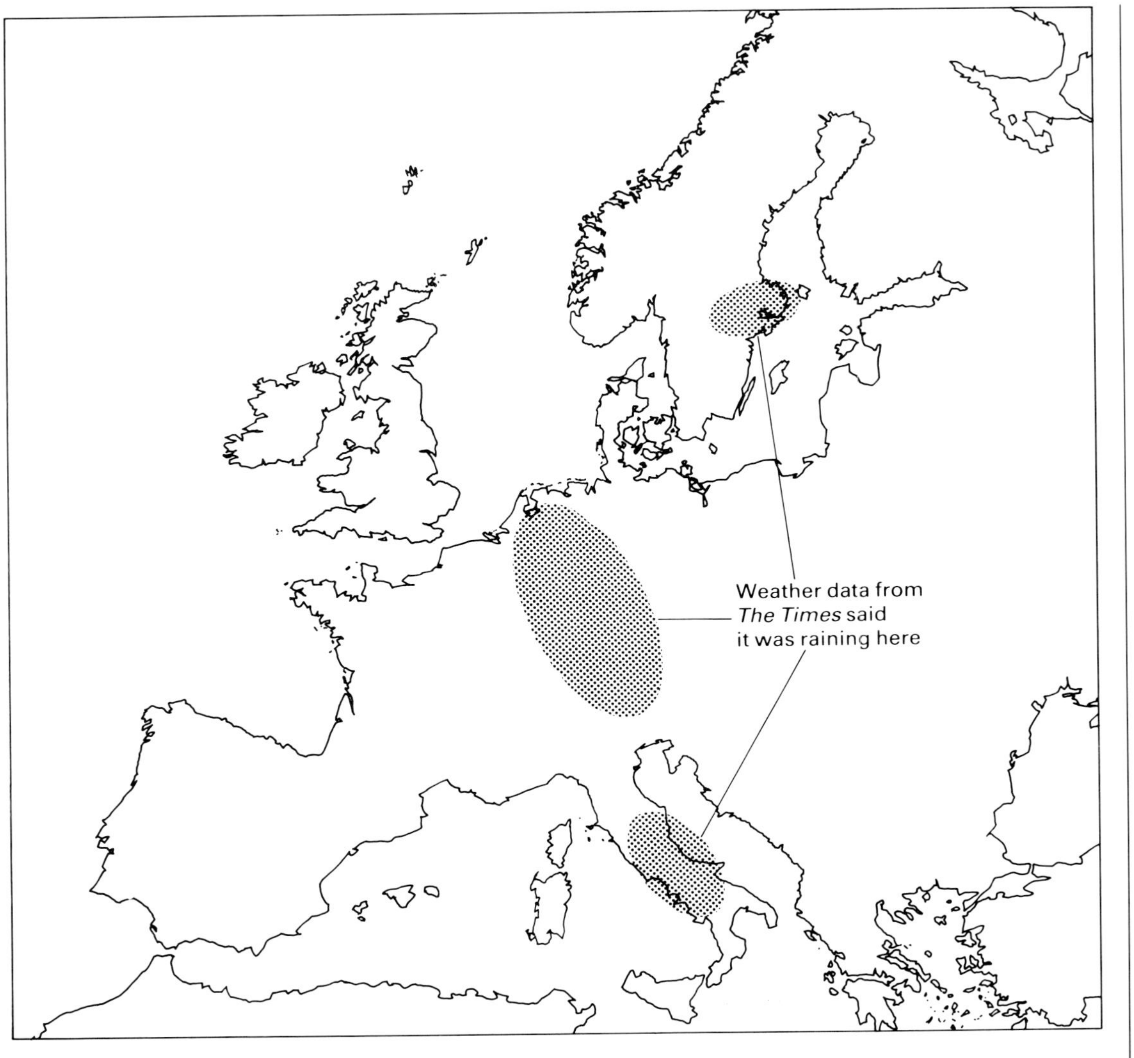

A simple rainfall map

3 You could try doing these activities each day for a week or so, and give yourself a score out of 10 each day – 10 out of 10 would mean that you had put the rain in the right places, and 0 out of 10 would mean you hadn't interpreted any of the rainy areas correctly!

2.7

Clouds and Rainfall on Weather Maps

Clouds and rainfall are shown on different kinds of weather map using different symbols.

On television and in newspapers, the weather is usually shown using pictograms, like the ones below.

- **Temperature** – the figure inside the circle indicates the temperature. Red numbers on a yellow background give positive figures, and black figures on a blue background show those below freezing. (15)
- **Sunshine** – the symbol shows sunny weather, with the number in the middle indicating temperature. (25)
- **Cloud** – this symbol means cloudy weather, with the clouds being thin or patchy.
- A black cloud means that there is a lot of cloud, with the weather being very dull.
- The sun and the cloud together show **sunny intervals** in between cloudy weather.
- **Rain** – the teardrop underneath a cloud means rain.
- **Showers with sunny intervals.**
- **Snow** is shown by the snowflake symbols.
- **Sleet** is shown by a snowflake and a raindrop.
- This symbol of a black cloud and yellow flash shows the possibility of **thunder and lightning**.
- The arrow on this symbol shows **wind direction**, and the number in the circle shows the **speed** of the wind in miles an hour. (30)
- FOG — **Fog** is shown by the word only

On weather maps produced by the Meteorological Office, however, weather information is shown in more detail, with a weather symbol for each weather station location. A weather map using these symbols is shown in the activities. Image 3 was taken on the same day as the weather data for this map was collected.

The weather symbols are made up of the following information:

Weather symbols

- Rain
- Drizzle
- Snow
- Rain and snow
- Hail shower
- Rain shower
- Snow shower
- Thunderstorms
- Fog
- Mist

Temperature (degrees Celsius)

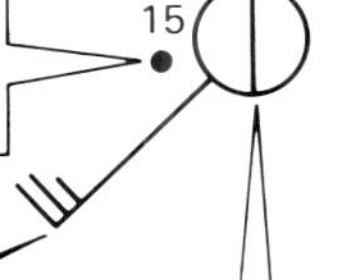

Wind

Wind direction shown by arrow 'shaft'

Wind speed shown by number of feathers

1 full feather = 10 knots
$\frac{1}{2}$ feather = 5 knots

Cloud symbols

Cloud cover is measured in eighths of cover, or *oktas*.

- Clear sky
- $\frac{1}{8}$ covered or less
- $\frac{2}{8}$ covered
- $\frac{3}{8}$ covered
- $\frac{4}{8}$ covered
- $\frac{5}{8}$ covered
- $\frac{6}{8}$ covered
- $\frac{7}{8}$ covered or more
- $\frac{8}{8}$ covered
- Sky obscured

2.7

ACTIVITIES

1 Look at the weather map below, and test yourself to see if you can work out what some of the symbols mean. You can do this activity in pairs if you want to.

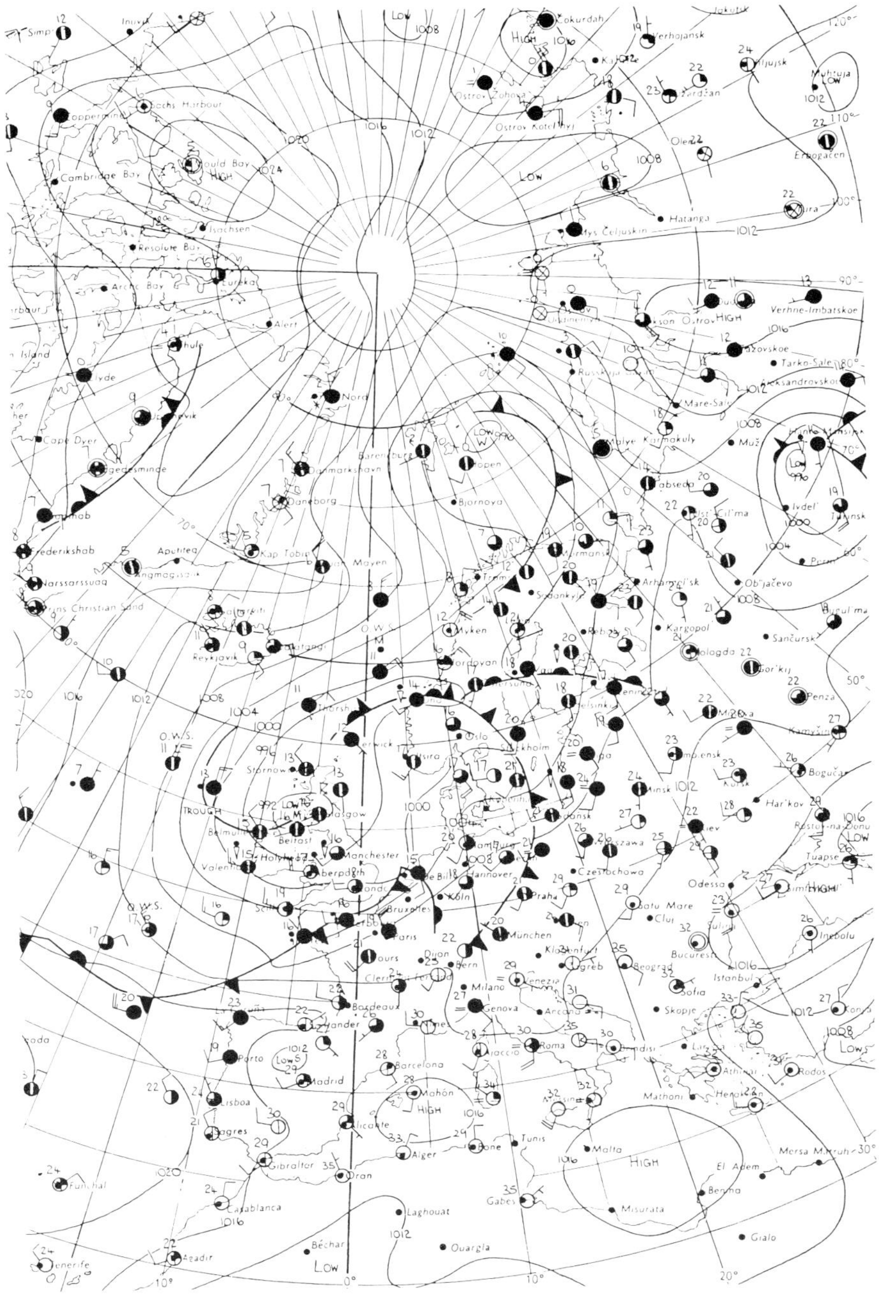

Weather map for 12.00 on 1 August 1979 (Images 3 and 4) (Crown copyright reserved)

2 Use the latest satellite images or any of the images provided in this pack, both infra-red and visible, to produce a television-style weather map to show where the cloudy areas of the UK and Europe are.

Unlike the weather station symbols, pictograms do not have to be placed over a particular location. They should, however, give a clear indication of what the weather is like over different parts of Europe at a glance.

2.8 Clouds through a Depression

Before you try this worksheet, it would be a good idea to have done some of the activities which look at depressions and anticyclones (3.1–3.7). When you look at the cross-section of a typical depression below, you will notice that there is a particular sequence of clouds.

ACTIVITIES

1 Look at the diagram below. It shows a cross-section through a depression. Fill in a copy of the table underneath the cross-section to describe the height and thickness of the clouds at each place, A–E. (Be careful since the scale is not linear.)

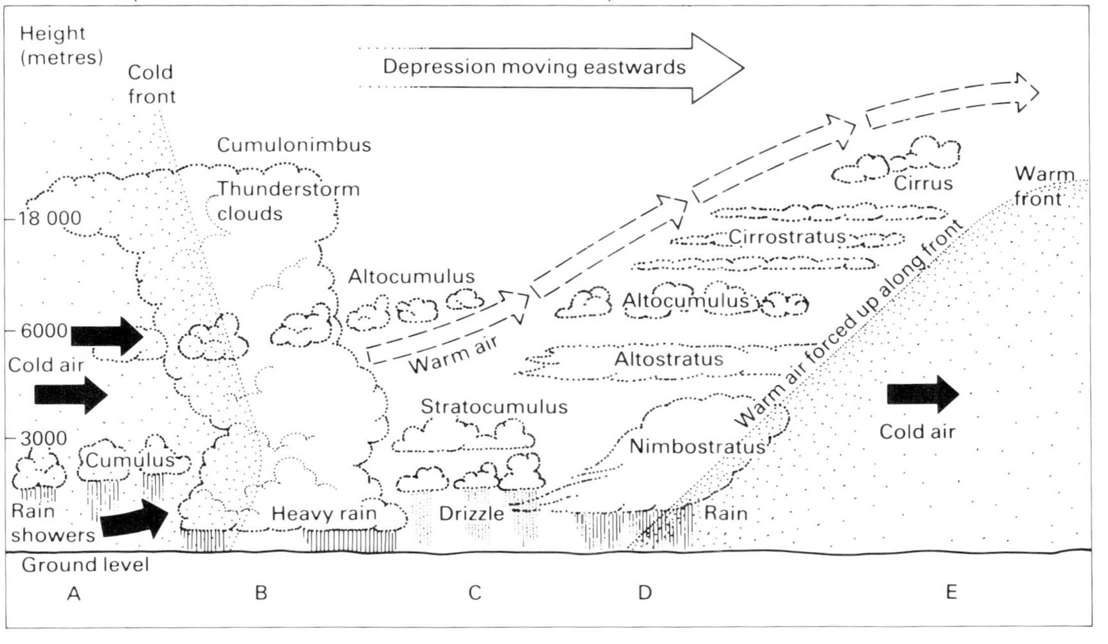

Place	A	B	C	D	E
Cloud name					
Cloud height					
Cloud thickness					

The cloud pattern along an occluded front is very much the same as the cloud pattern along the cold front, though the band of cloud is often wider.

2 Look at Image 5 alongside the sketch on page 32. It shows the different types of cloud that you will often find in a depression, and should help you to interpret your own images.

Can you spot this typical sequence of clouds in a depression on a satellite image?

2.8

If you are doing this activity with your own current infra-red and visible images you will need a copy of a weather map from a newspaper or other source. You can also do this activity using Image 9, in which case you will need the weather map in Appendix 5.

On *two* copies of the right base map, mark the position of the fronts that make up the depression. On one of the maps, use the visible image to mark on the thickness of the clouds. On the second map, use the infra-red image to mark on the height of the top of the clouds.

Can you spot any relationship between the fronts on the map and cloud height and thickness?

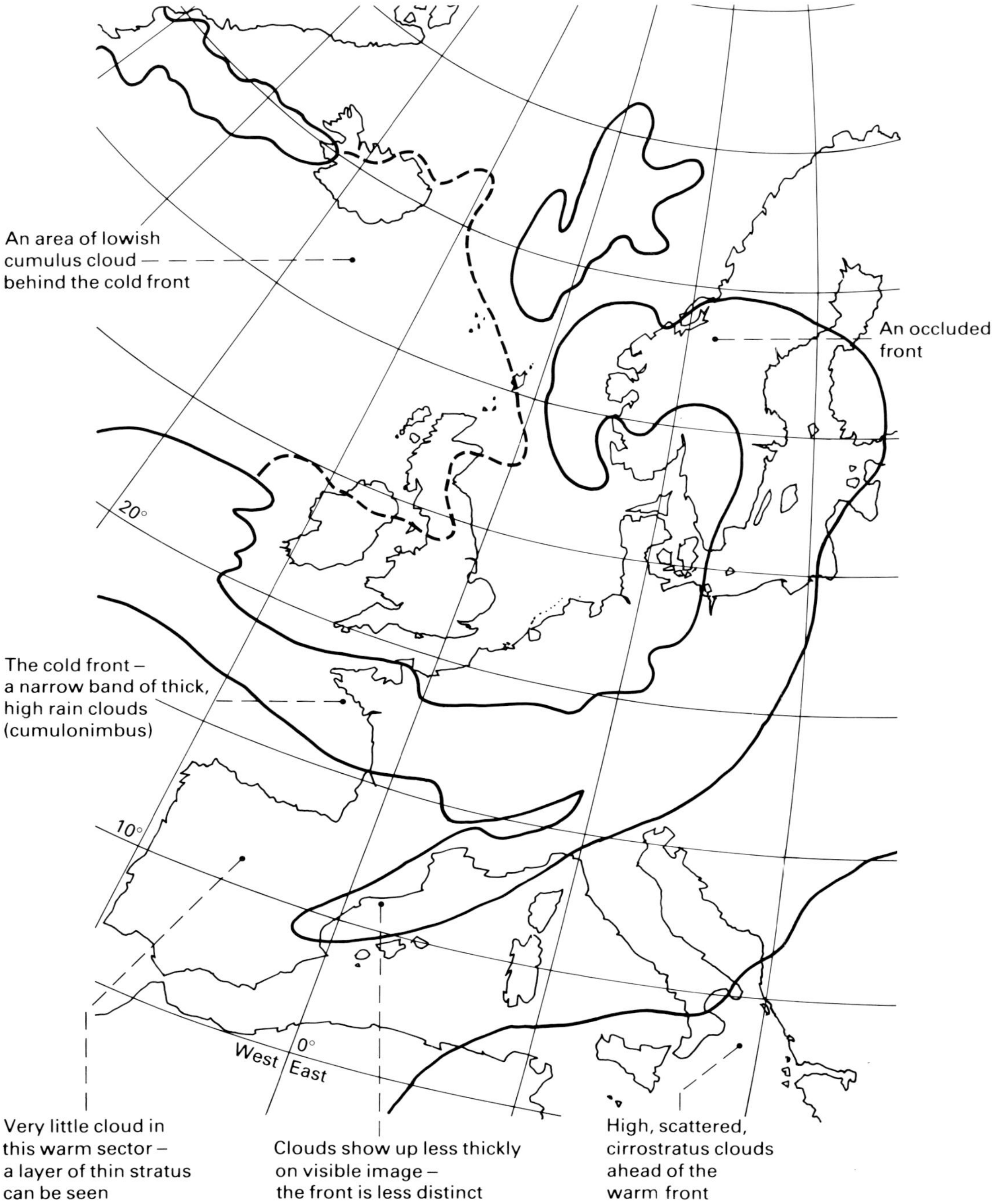

Clouds in a depression on Image 5

2.9 Rainfall, Rivers and Flooding

Flooding is a major hazard on a world scale, and also, though to a much lesser extent, in the UK.

Floods happen on rivers for a number of reasons, and these are summarised on the diagram below. Basically, as you can see from the diagram, if the ground is saturated or almost saturated with water, then during heavy rainfall nearly all the water that falls goes straight into the rivers. If the rivers are already full of water, floods are likely to occur.

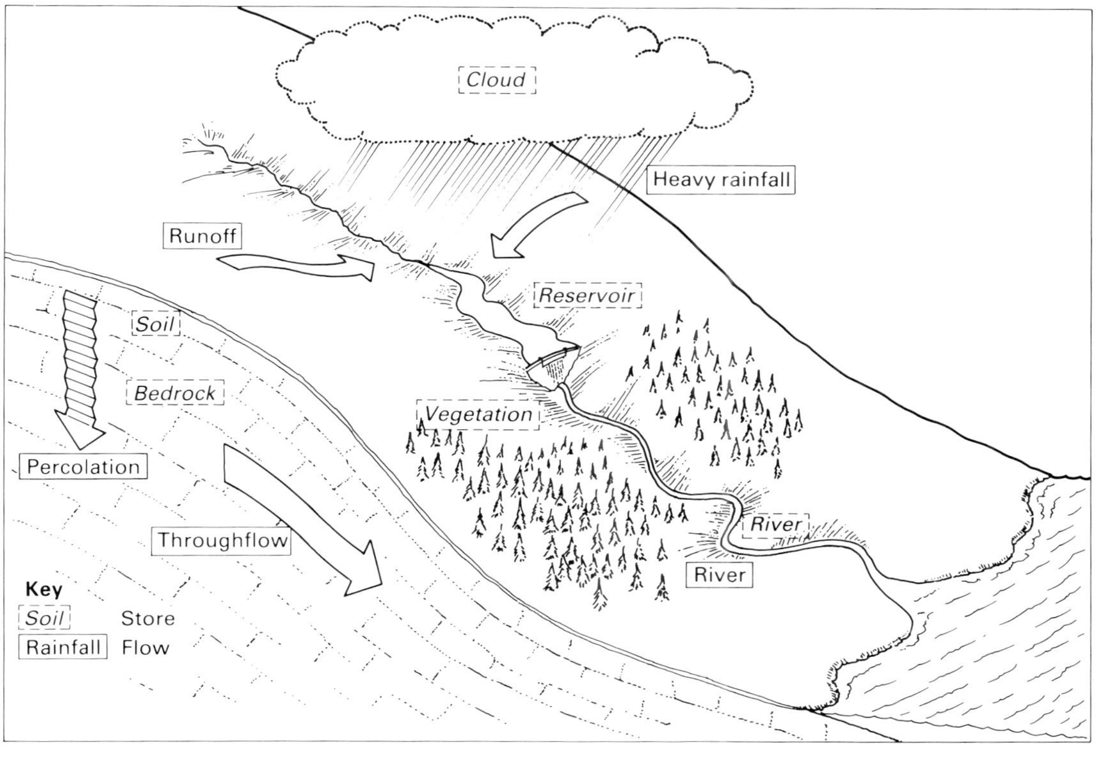

The water cycle – stores and flows

Once all the stores are full, flooding is likely.

In England and Wales, the water authorities are responsible for the management of rivers, and this includes giving flood warnings for the rivers under their control. The problem they have, of course, is to predict accurately when a river will flood. The water authorities will know from weather forecasts when rainfall is on its way, but a satellite image won't tell them exactly how much rain will fall in a given area. They need to know this to issue a flood warning in time, as it is obviously no good putting out a flood warning every time it rains – people in areas liable to flooding don't want to have to go to the bother of moving out of their homes if there is no real danger of a flood!

Instead of satellite images, therefore, *remote sensing images* produced by radar are used. A radar image is able to give an exact indication of the intensity of rainfall in millimetres per hour, and of where that rain is falling. This gives the water authorities the ability to predict which rivers will flood and where.

This is called *nowcasting* in the business, because radar images can only be used for short-term predictions. They also cover only certain areas of the UK.

2.9

ACTIVITY

In this decision-making activity you have to imagine that you are in charge of issuing flood warnings for the Welsh Water Authority on one occasion – 27 December 1979. This date was a crisis time because there had been a lot of rain, the ground was wet and rivers were relatively full. You should do this activity in a group of three or four.

You should study the radar images available for this date (shown on page 36), and try to work out which rivers in Wales are most likely to flood. You should issue flood warnings if the rainfall in an area is over 4 mm an hour. You can assume that the pattern of rainfall has been very similar for the past five hours, and will continue for the next five hours before dying out.

On the base map of drainage basins in Wales, given on page 37, shade the rivers where you would expect flooding to take place and for which you would issue flood warnings.

After you have done this, study the newspaper extracts for the same date, which give you some details of where flooding actually took place. Did you predict the flood accurately?

Note: real radar images are in colour. This has been simplified to make it easier to interpret.

HUGE AREAS of Wales were awash under the worst floodwaters for 20 years early today. Two pensioners were swept to their deaths, towns and villages were cut off, and hundreds of families were made homeless.

A massive squad of emergency service workers was also standing by early today to cope with new floodwater chaos they were fearing from the early morning high tides which hit Wales between 9 p.m. and 2 a.m.

Reception centres were set up throughout the South Wales valleys for more than 1000 people evacuated from their homes, and in **Cardiff** the city's entire bus fleet was taken over to cope with the homeless.

The worst hit areas were industrial South Wales and Cardiff itself but in **Brecon** troops and airmen joined emergency services in a massive rescue operation as parts of the market town disappeared under five feet of floodwater.

The weathermen said last night that rainfall in South Wales was the heaviest in 10 years – and they predicted more to come.

The two pensioners who died were drowned when a flooded brook burst its banks and sent a tidal wave 10ft high crashing into their cottage at Rhydycar, **Merthyr**.

Elsewhere in the Merthyr area scores of families were evacuated as the torrential rain caused widespread chaos.

The **Aberfan** Community Centre was one of half-a-dozen centres in the valleys which were used to accommodate evacuated families.

Police said the **Rhondda** area was virtually cut off with floods up to eight feet deep in places, and in **Pontypridd** the town's shopping centre was flooded.

Dozens of people had to be rescued by **West Glamorgan** firemen as scores of homes were flooded by the heavy rain.

Gales reached 60 m.p.h. and an inch-and-a-half of rain fell in the **Swansea** area in 12 hours.

Flooding was worsened by high tides early yesterday, and more flooding was feared late last night as another high tide was due.

A newly-wed couple living in a riverside flat at **Newcastle Emlyn** were made homeless as floodwater and sewer water poured knee-deep into the living-room and bedroom of their home.

The couple, Mr. and Mrs. Gary Francis, only moved into the ground-floor flat of a house alongside the Adpar river bridge over the River Teifi in October.

At **Neath** floodwater broke down an eight-inch thick wall. At the local technical college water flowed from the grounds on to the main Neath–Pontardawe road, closing it for most of the day.

In **Dyfed** dozens of main roads were flooded and electricity and telephone services badly hit.

2.9

Trees down

At **Whitemill**, near **Carmarthen**, the village public-house was flooded and firemen had to pump gallons of water out. The gales brought down trees in many areas and police said scores of motorists had abandoned their cars because of electrical faults.

Farmers in the **Towy Valley** moved hundreds of sheep to higher ground as the river burst its banks. Hundreds of acres of farmland were under water last night.

The main A40 in West Wales was blocked by floodwaters in at least six places between Carmarthen and **Brecon**. Floodwater roared across the road at **Abergwili**, near Carmarthen, and at **Rhosmaen**, **Llandeilo** and **Llandovery**.

Other roads blocks included the A470 **Cardiff** to Brecon and the A487 at **Aberystwyth**.

Emergency services throughout Wales had to cope with the floods, which were caused by up to three inches of overnight rain.

In **Mid-Wales** the town of Brecon was completely cut off, and Aberystwyth was barely accessible as floodwater continued to rise during the day.

'Flooded road' signs were placed by police on the highway alongside the Dovey Bridge at **Machynlleth**, but by mid-afternoon cars were still managing to get through.

Many petrol stations in Dyfed could feel the after-effects of the torrential rain which flooded a major storage depot in Haverfordwest.

At one time the Shell depot on the outskirts of the town was under four feet of water and the 1,150,000 litres of fuel was marooned because the service pumps were almost totally under water.

It's no-go for motorists

ROADS throughout Wales were awash after yesterday's floods – and police advised drivers not to go out unless their journeys were essential.

An Automobile Association spokesman said conditions for motorists had not been as bad since the snow of January and February this year.

All 25 telephone lines to the A.A. in **Cardiff** were jammed as hundreds of drivers, many of whom were stranded in their vehicles, called for assistance, and the spokesman said, 'Our biggest trouble is getting to them.'

In Cardiff both the River Ely and Taff burst their banks. Llandaff Bridge was closed to traffic and Cathedral Road became impassable.

Several roads in the **Merthyr Tydfil** area were blocked. The A470 was flooded at **Pentrebach** to the south and between Merthyr and Brecon to the north; the A465 was impassable at **Cefn Coed**.

A police spokesman in Mid Glamorgan said he had lost count of the number of roads closed. The A4058 was impassable and the Rhigos mountain road was blocked.

The main **Pontypridd** to **Treherbert** road in the Rhondda was flooded in many places and the A474 was flooded north of **Neath**.

Only two roads were open in **Gwent** – the A4042 between Newport and Cwmbran, and the A48 Newport to Chepstow road. **Abergavenny** and **Usk** were cut off and every other town was flooded.

The M4 remained open but high-sided vehicles were stopped from crossing the Severn Bridge.

Many roads were flooded in **Dyfed** and **Powys**. The A40 was blocked at several places including **Crickhowell, Brecon, Llandeilo** and **Abergwili**. At **Whitland** the road was three feet under water.

The Llandovery to Llangadog road was cut as was the A484 between Carmarthen and Cardigan. The A474 was blocked between Ammanford and Ty Croes.

The A470 was cut at Brecon and Builth Wells, and was closed between Builth and Rhayader.

The only road out of **Builth** was the A483 to the west – the stretch north of the town was blocked.

Conditions in North Wales were not as bad as elsewhere in the Principality. The A55 was flooded between Abergele and St. Asaph at **Bodelwyddan** and the B5106 was blocked between Llanrwst and Trefriw.

2.9

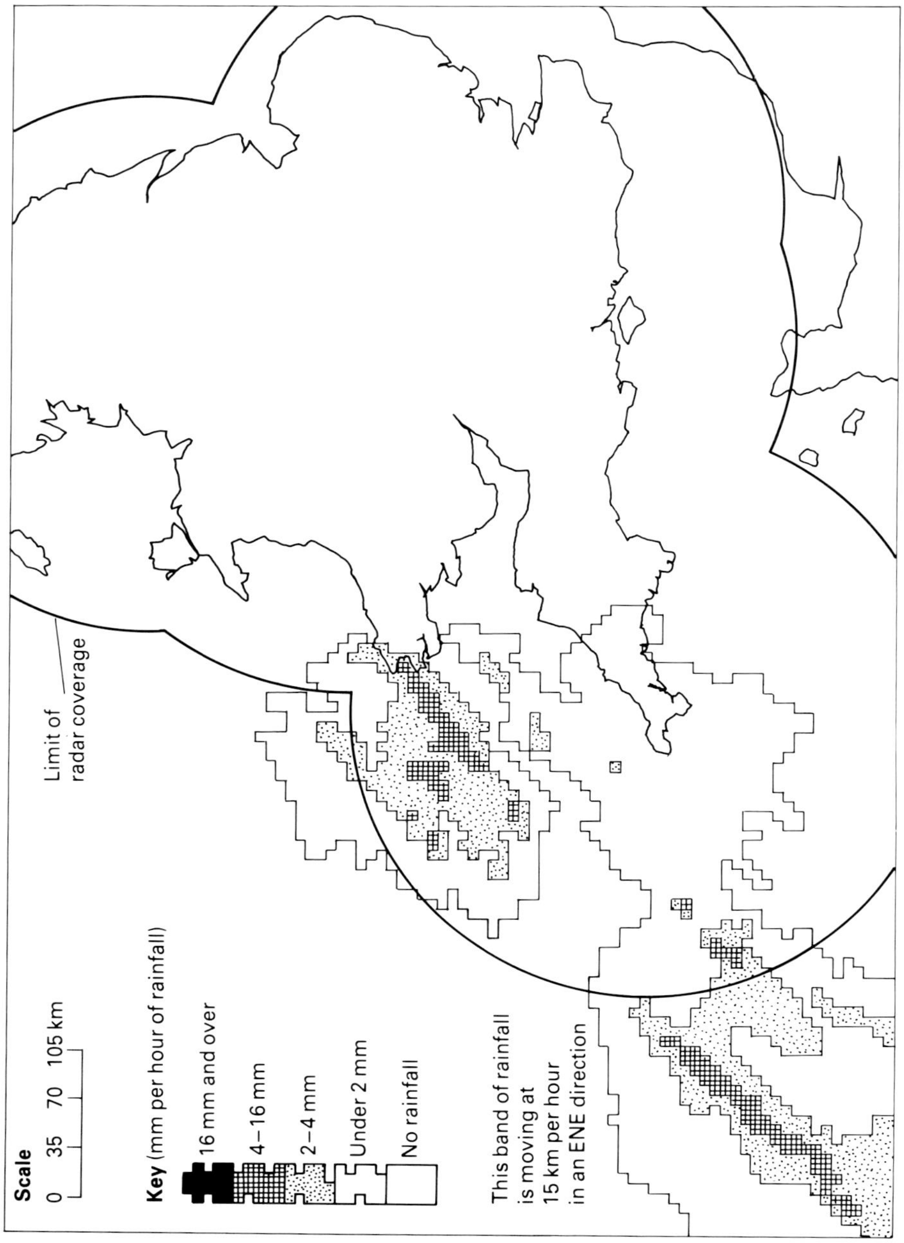

Simplified radar image for 01.00 on 27 December 1979

2.9

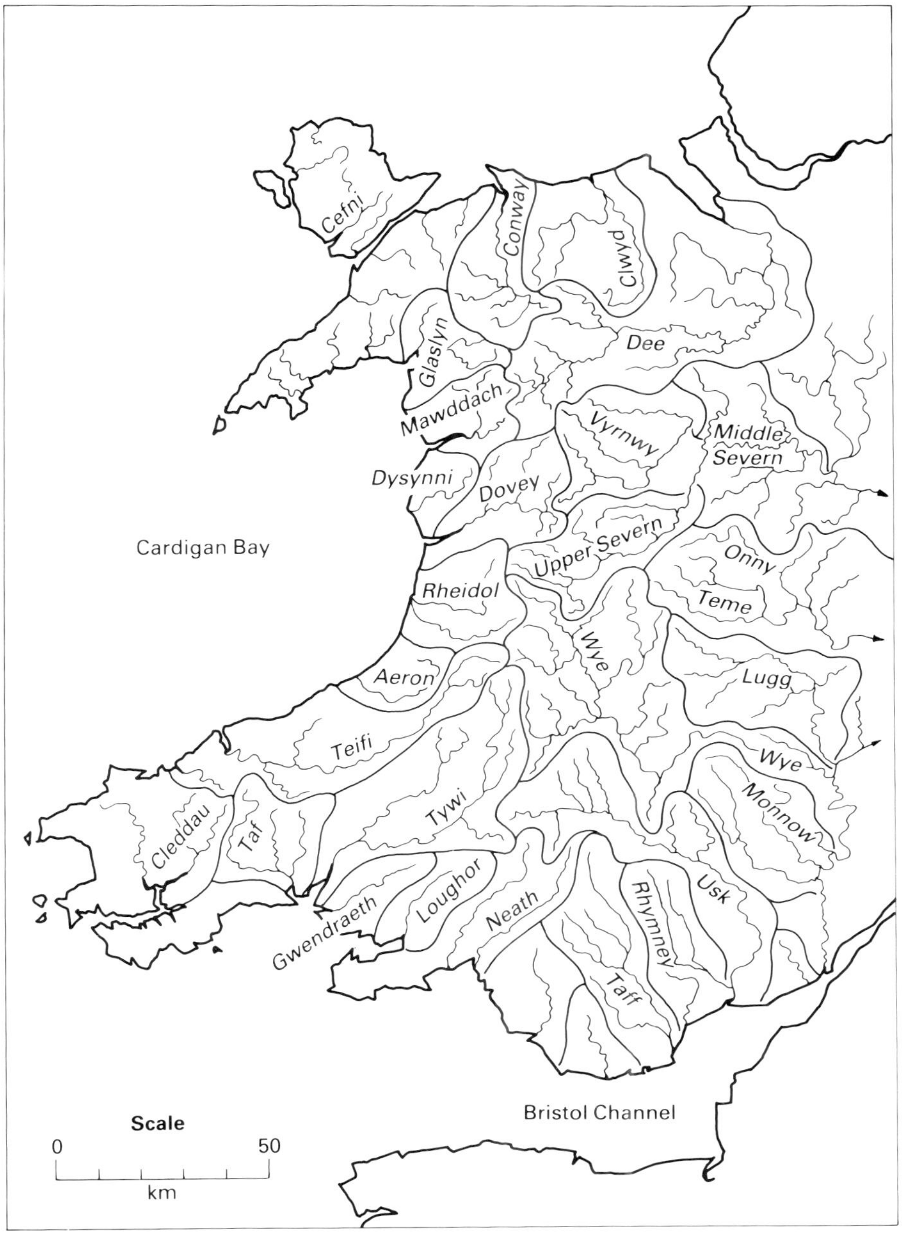

Drainage basins in Wales

OBJECTIVES CHECKLIST

2.1 What the Images Mean – Visible

Tick here

Do you

- understand what a visible image is? ☐

Can you

- take measurements about clouds from a diagram? ☐
- find clouds of different thicknesses on a satellite image? ☐
- transfer information from a satellite image to a map? ☐
- use an atlas? ☐

2.2 What the Images Mean – Infra-red

Do you

- understand what an infra-red image is? ☐

Can you

- use a line graph to work out cloud temperatures? ☐
- use an infra-red satellite image to work out cloud top height? ☐
- transfer information from a satellite image to a map? ☐
- use an atlas? ☐

2.3 How are Clouds Formed? (1)

Do you

- know and understand how clouds are formed? ☐
- know some names of clouds? ☐

Can you

- work out what different kinds of cloud will look like on a satellite image? ☐
- describe how you feel about different types of cloud and the weather they bring? ☐

2.4 How are Clouds Formed? (2)

Do you

- know about and understand the different ways in which rainfall happens? ☐
- know the names of the different kinds of rainfall? ☐

Can you

- link together information from visible and infra-red satellite images to find different kinds of cloud? ☐
- link written material with different kinds of cloud? ☐

2.5

Clouds for Real

Can you

- describe clouds? ☐
- say what the clouds are like over your school, using a satellite image? ☐
- keep a cloud diary? ☐
- make a poster to show what clouds look like on satellite images? ☐

2.6

What about Rainfall?

Can you

- use a satellite image to work out where it should be raining? ☐
- use information from Prestel/Ceefax to test your prediction of rainfall? ☐
- use newspaper information to make a map to show where it has been raining? ☐
- correlate (compare) your satellite map with actual areas of rainfall in Europe? ☐

2.7

Clouds and Rainfall on Weather Maps

Can you

- interpret both media and Meteorological Office weather maps? ☐
- draw a weather map using pictograms? ☐

2.8

Clouds through a Depression

Can you

- take information from a diagram and record it in a suitable form? ☐
- link information from infra-red and visible images to identify different cloud types through a depression? ☐

2.9

Rainfall, Rivers and Flooding

Do you

- understand why rivers can flood? ☐

Can you

- interpret radar images? ☐
- use radar images to predict flooding? ☐
- use newspaper weather information? ☐
- work in a group? ☐

3.1

Depressions and Anticyclones

If you listen to the weather forecast regularly, then you will have heard about *depressions* and *anticyclones* lots of times. Both of these things are very large-scale weather features, often hundreds of kilometres across. Depressions are areas of *low pressure* (sometimes called cyclones if the pressure is very low), and anticyclones are areas of *high pressure.*

At times of low pressure the air is usually rising. As it rises it cools and condenses and clouds form. Low pressure is associated with depressions, and depressions bring with them clouds and usually rainfall.

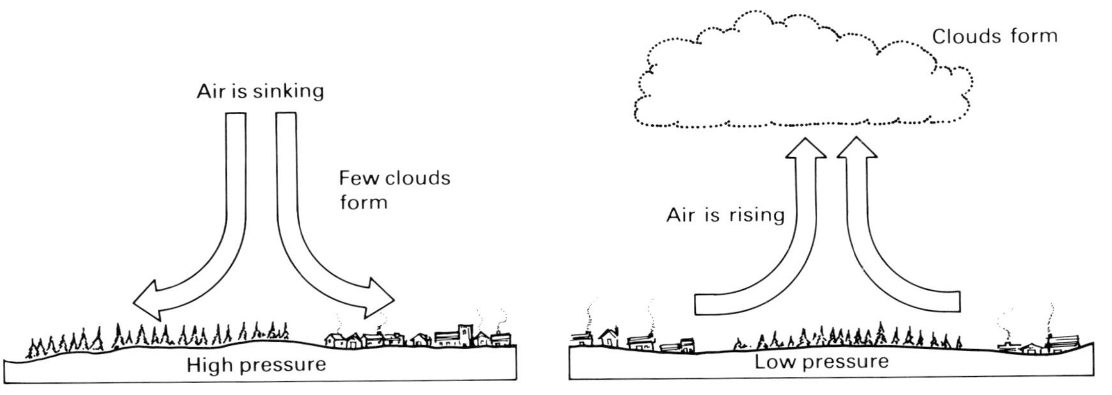

On the other hand, high pressure means that air is falling. As it falls, it gets warmer, and therefore it is able to hold more moisture. As a result, few clouds form and high pressure is associated with cloudless, dry weather.

Pressure is also responsible for winds. The diagram below shows that winds will blow from high pressure areas to low pressure ones. You can almost imagine the low pressure areas 'sucking' the wind from the high pressure ones as the air is rising. The lower the pressure, the stronger the wind.

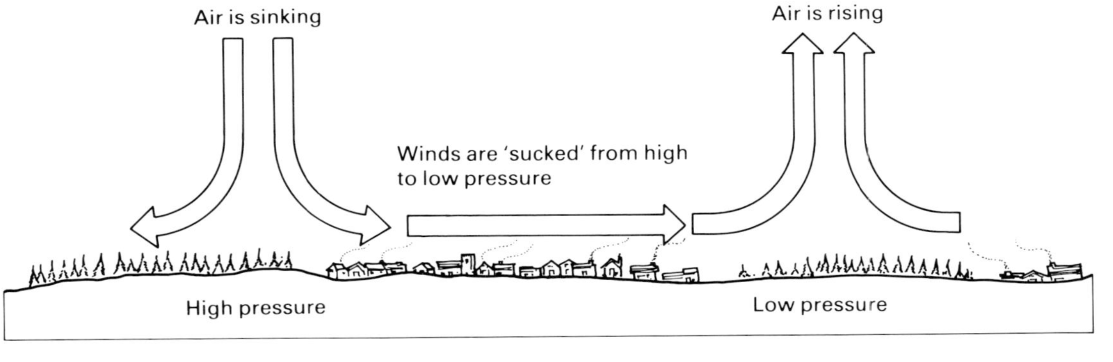

Pressure is shown on a weather map by *isobars*. These are lines of equal pressure. Pressure is measured in *millibars*, and low pressure areas such as depressions usually have figures of around about 980–1010. Anticyclones or high pressure areas have figures of 1020 or more.

Because of the rotation of the earth, the winds don't blow straight from high to low pressure areas. Instead they tend to spiral into low pressure centres – a bit like water going down the plughole in a bath! The lower the pressure is and the closer the isobars are together on a map, the stronger the winds will be.

3.1

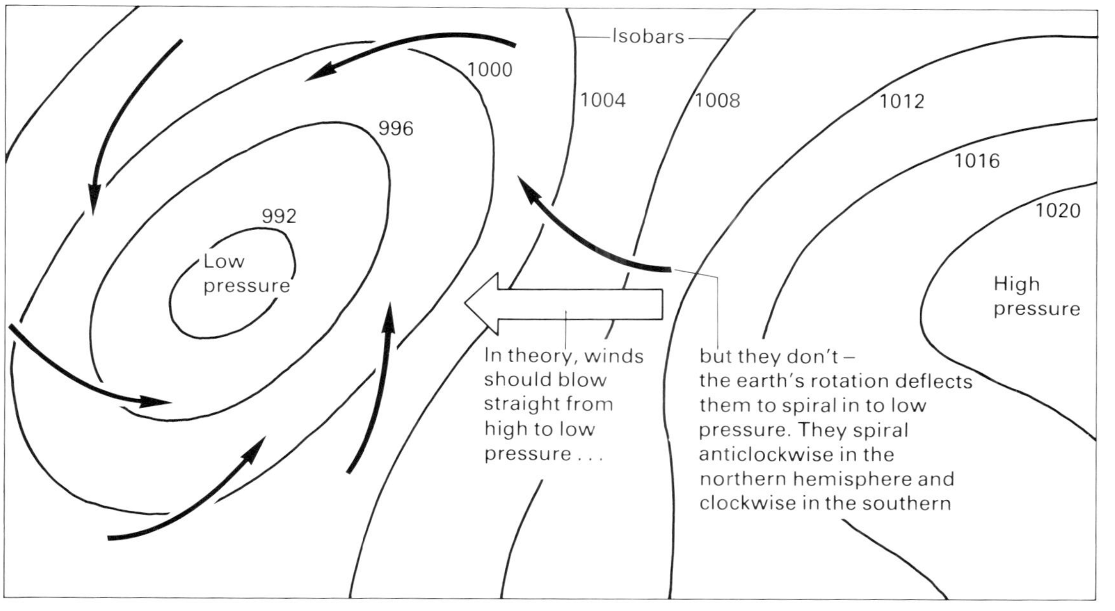

ACTIVITIES

1 Look at Images 2 and 3. You should be able to tell easily which parts of the images have high pressure, and which have lower pressure.

Look at the current satellite image and say whether there is high or low pressure over the UK. Check your answer with information from a media weather map.

2 Look at the pressure map on page 42. It shows the pressure over parts of Europe at about the time Image 5 was taken.

Use the hints on the map to help you to finish drawing in the isobars. When you have done this, mark the direction of the winds on the map as well, and also note whether you think the winds at the time were strong or gentle.

Some of the isobars have been done for you. You should join together all the places with the same pressure. You should also put isobars in for every 4 millibars.

Sometimes you may not have enough points on the map with every value. This doesn't matter. You just put these lines in between the others.

3 Imagine that you are in charge of a ship which has the task of supplying North Sea oil rigs.

You have to plan your sailings carefully, because if you sail in high winds, the waves these winds cause will make it dangerous for you to get close to the rigs.

Look at each of the weather maps on pages 42–3. For each one, describe the direction and likely strength of wind over the North Sea. On which of the days do you think you would be able to supply the rigs?

3.1

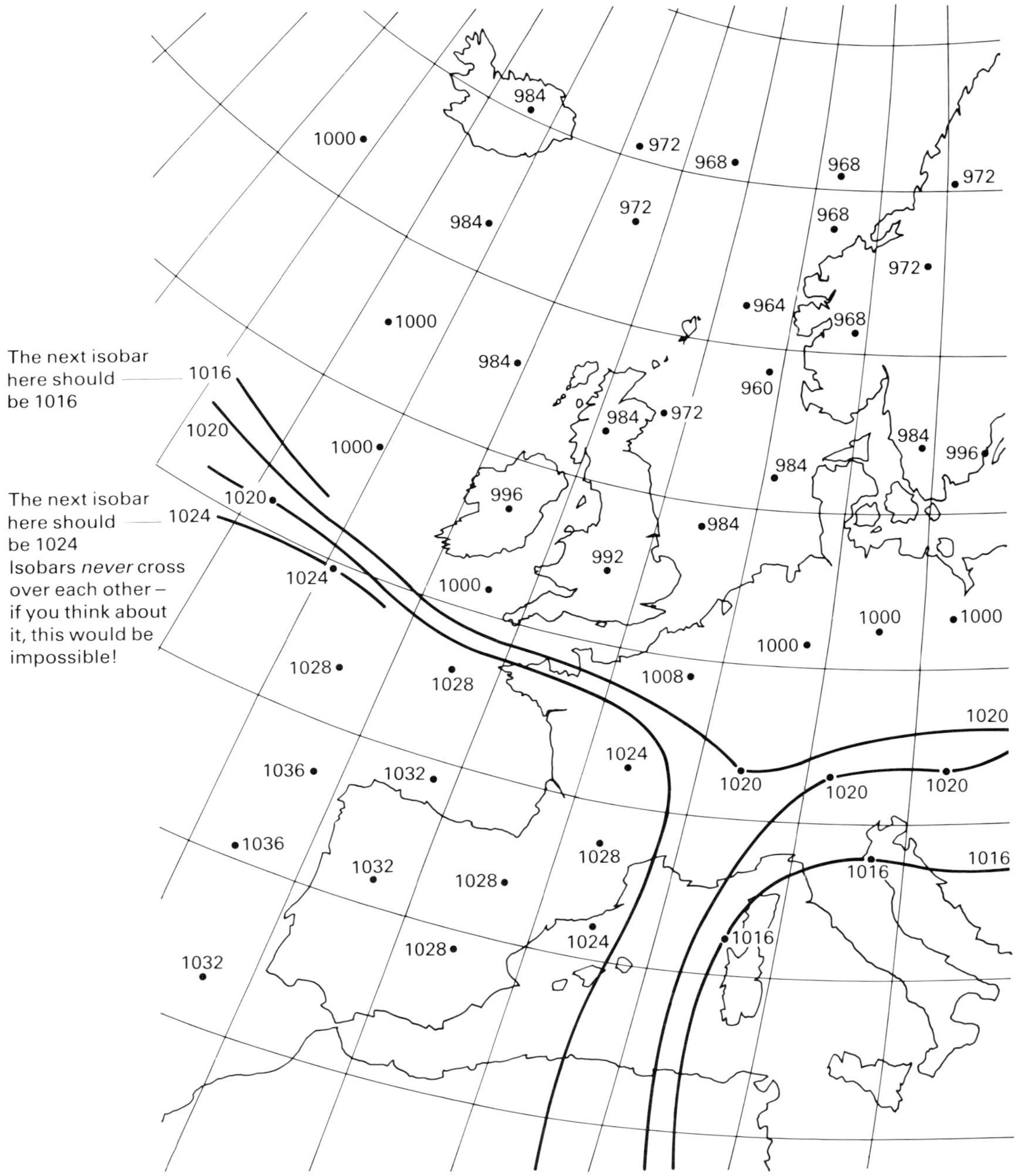

Pressure map for Activity 2 (for 12.00 on 13 January 1984)

NOON: 2 MARCH

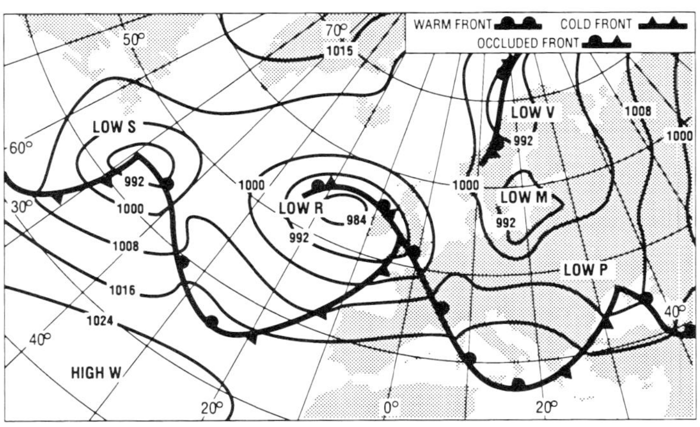

Weather maps for Activity 3 (continue on page 43)

3.1

NOON: 4 MARCH

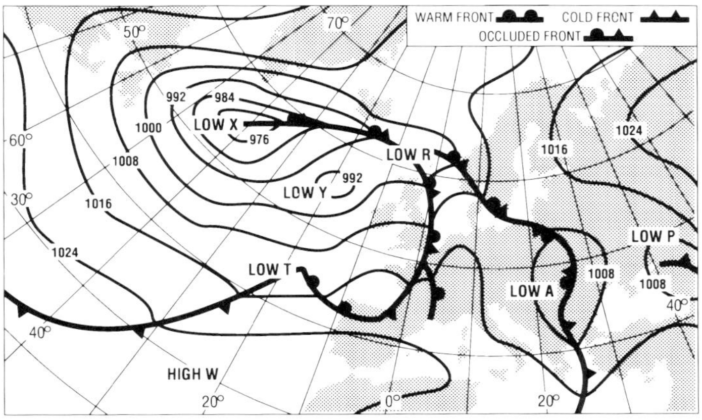

NOON: 6 MARCH

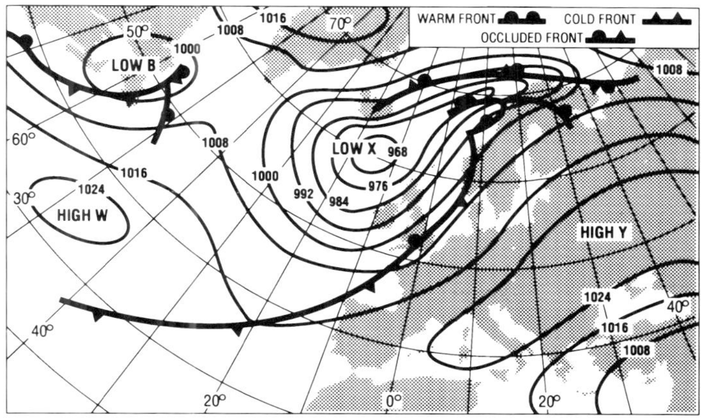

NOON: 13 MARCH

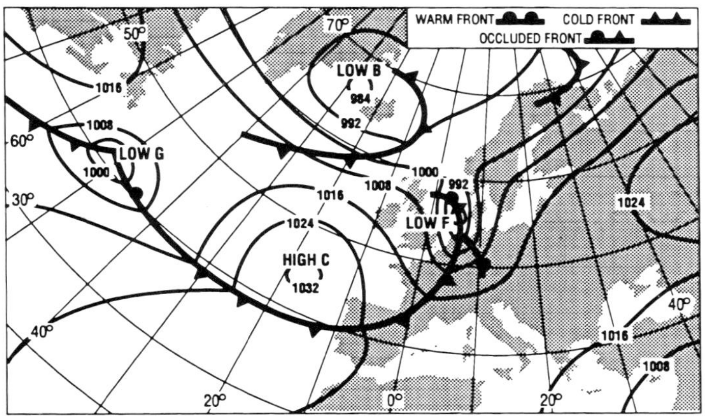

NOON: 15 MARCH

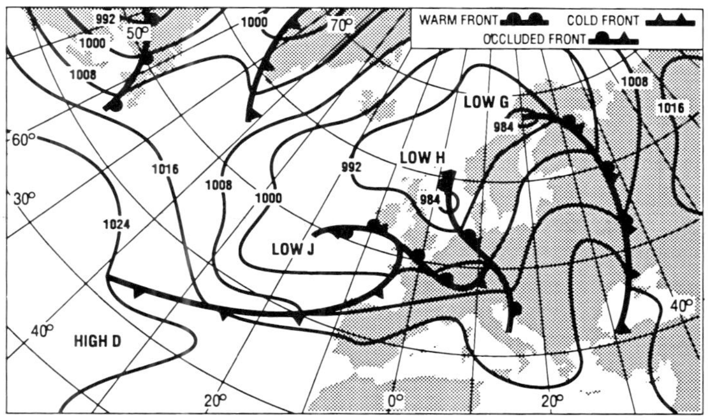

Weather maps for Activity 3 (continued)

3.2

Depressions and Air Masses

A depression is a centre of low pressure. Because of this winds are 'sucked' into this centre from surrounding areas of high pressure. Due to the fact that the earth is rotating, this wind spirals into the depression and the result is that a depression is usually seen on a satellite image as a large swirl of cloud.

Look at Image 5 alongside the sketch below. They both show the same area. This is a typical depression, and it should help you to recognise other depressions on live satellite images.

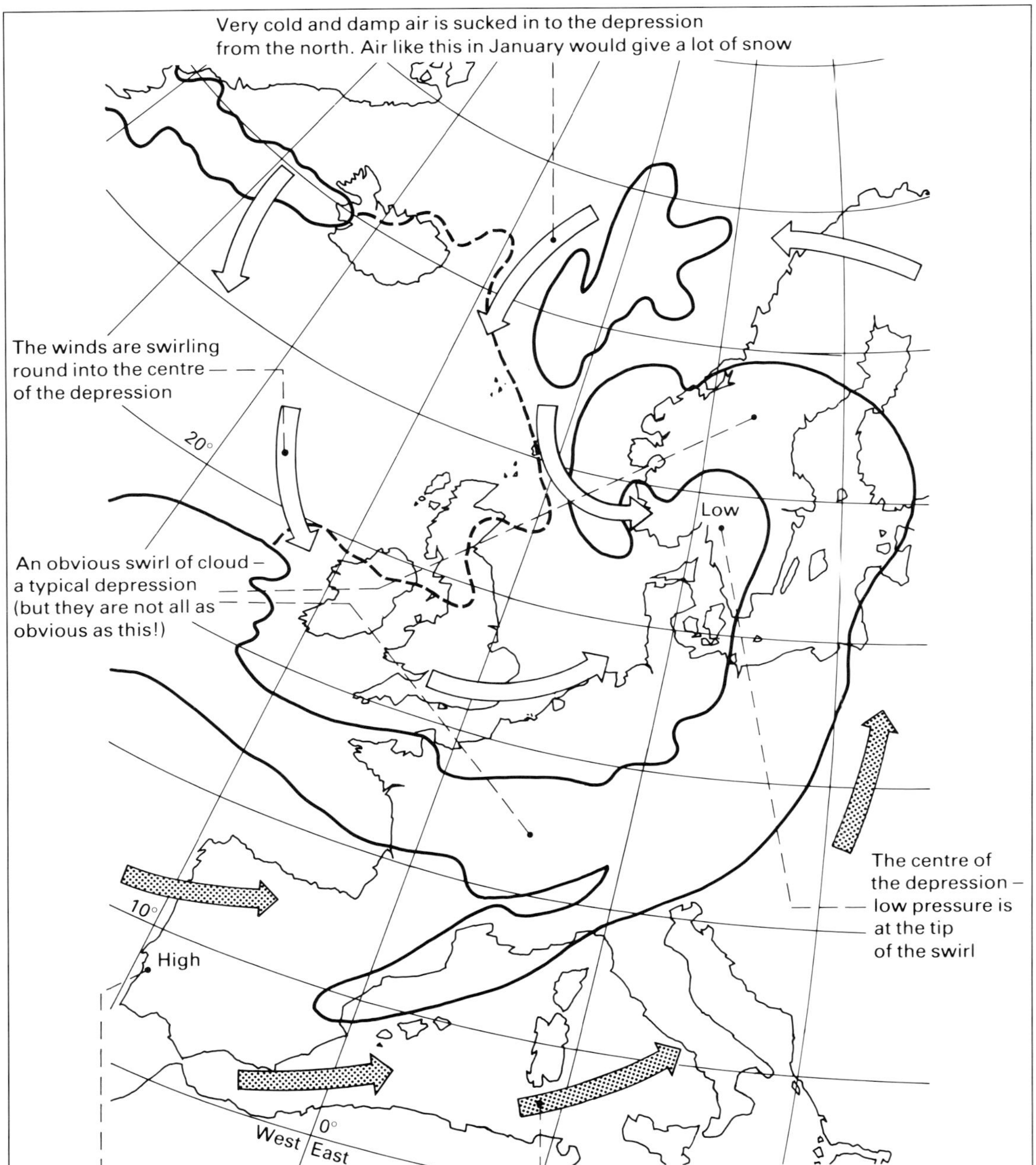

Winds in a depression on Image 5

3.2

The winds that are sucked into the depression can come from a number of areas. Where they come from will determine what kind of weather they bring.

The winds will come from an *air mass* that is located over a part of the earth. An air mass is exactly what its name suggests it is – a large mass or body of air covering a particular area. The map below shows which air masses will affect the weather in the UK, and what weather each one is likely to bring.

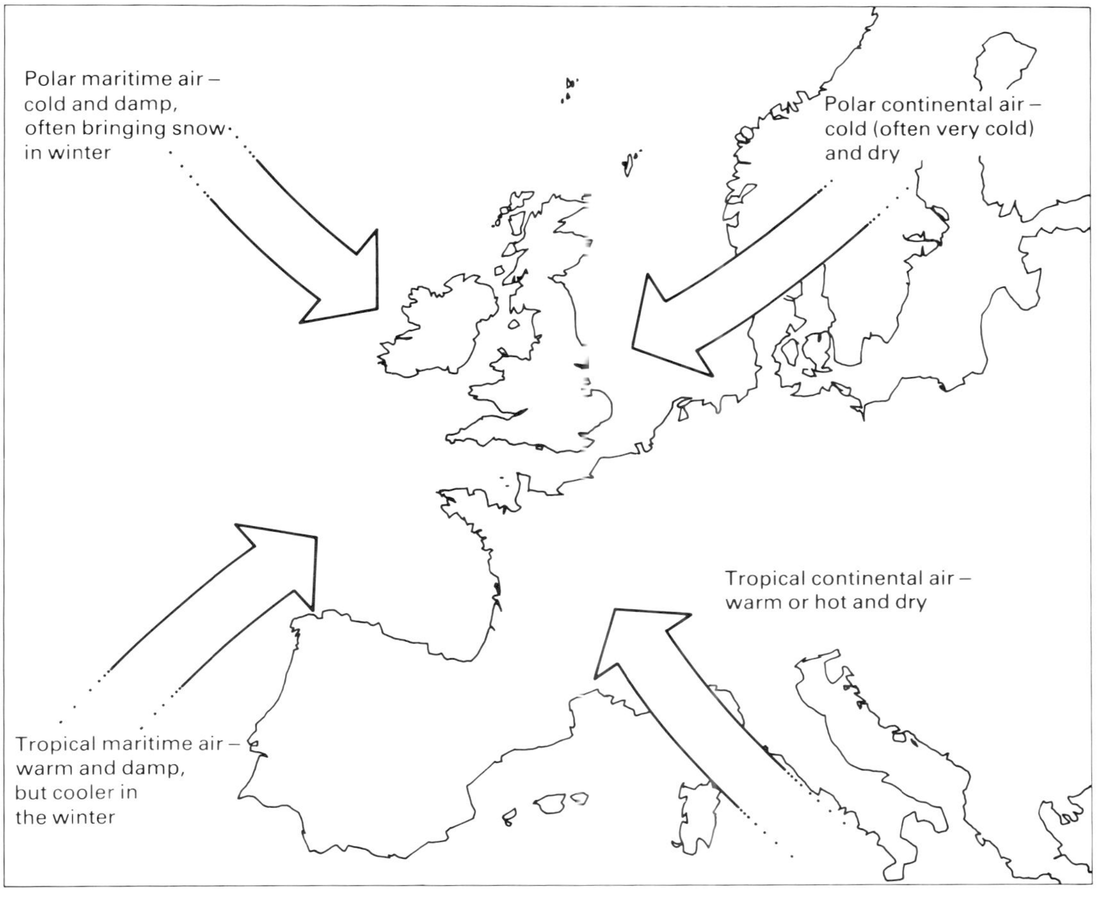

ACTIVITIES

1 Look at all the satellite images which go with this pack. Make a list of the ones that you can spot a depression on.

2 Either use your own satellite image together with a media weather map for the same day, or Image 9 and the weather map in Appendix 5, for this activity.

a Firstly, find the depression on the satellite image. On a copy of the right base map, mark the centre of the depression with the word 'Low' or letter L. This shows it is an area of low pressure. You can check your answer on a weather map.

b Then mark on your map the swirl of cloud that marks the depression.

3.2

c Now mark on the winds that are spiralling into the depression. Remember that they will spiral in an anticlockwise direction.

d Next you should try to work out whether the winds you have marked on will be cold or warm. Don't forget that in winter polar air masses are cold or very cold, whilst the tropical ones are cool. In the summer the polar air masses are cool or cold, and the tropical ones are warm or hot.

You could mark the different winds on your map with different colours to show their relative temperatures – red for warm and blue for cold.

e Finally, look at the media weather map, and find the areas of high pressure that the winds are coming from. You can mark these on your map with the word 'High' or letter H.

You could try to find out the temperatures of the areas the winds are coming from. For example, if there is high pressure over Spain, try to find out the temperature of Madrid.

3 Depressions are always on the move! If you used your own satellite image for Activity 2, then look at another image for later on the same day. Mark the centre of the depression on again, with the word 'Low', and then work out how far the depression has moved, and which direction it has moved in. You could also calculate its speed.

If you used Image 9, then you can still do this activity – Image 10 shows the same depression at a later time. You can use this to calculate its movement.

4 The different air masses bring with them different kinds of weather. Look at the start of the newspaper article below.

a What kind of air mass do you think was over the UK at the time?
b Finish the newspaper report, outlining some of the problems this air mass would cause old people.

'Hundreds of old people at risk', says Labour leader

As the bitterly cold east wind continues hundreds of old people are shivering in their homes, according to the leader of the Labour Party. 'The government should take action straight away, and give old age pensioners an extra heating allowance on top of their pension.

3.3

Fronts

Why is there nearly always a swirl of cloud in a depression?

The reason is that the cloud marks the boundary line, or *front*, between two different air masses. The air from two air masses, say a polar maritime air mass and a tropical maritime air mass, doesn't mix very easily because it is at different temperatures and densities in the different masses.

Along the boundary line which separates the two air masses, the colder air from the polar area will cool the warmer tropical air it comes into contact with. As it does this, the water vapour in the warm air condenses to form clouds. The warmer air is also lighter than the colder air, and will tend to rise up over it. As it rises, it cools and the water vapour condenses. The boundary line, or front, is therefore marked on a satellite image by a band of cloud. Often, it rains along this front line.

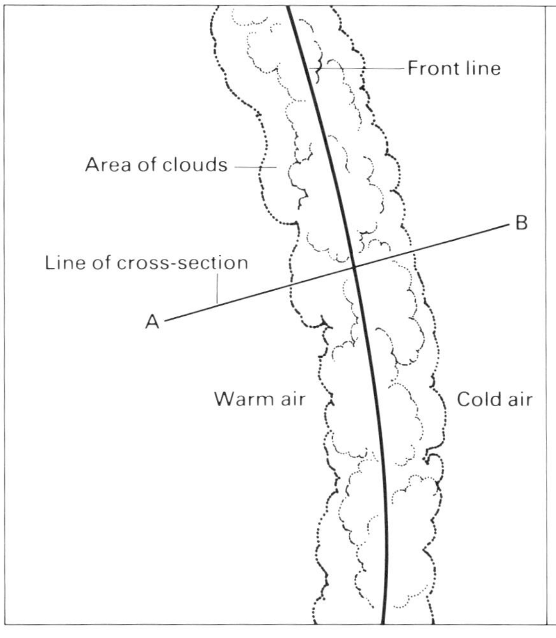

Map

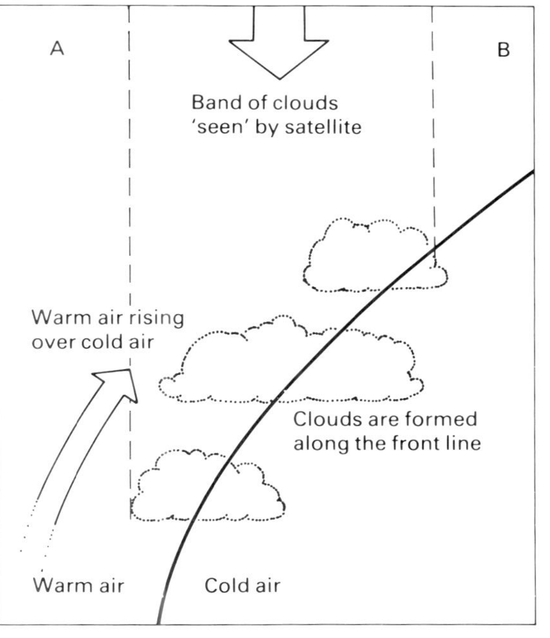

Cross-section

The front between two air masses is not a vertical line, nor is it as steep as the diagram above suggests. The angle of the front with the ground is usually between 5° and 20°.

It is interesting to note that the name 'front' was given to the boundary line between two air masses by meteorologists at the time of the First World War. They compared the air masses to two armies, both battling to control the weather over the UK. The front line divided the two air masses, just as the front line divided armies on the ground.

A depression nearly always has a number of fronts associated with it. Image 5 shows a number of fronts, and you should look at this alongside the sketch on page 48. This should help you to recognise fronts on the images you receive.

3.3

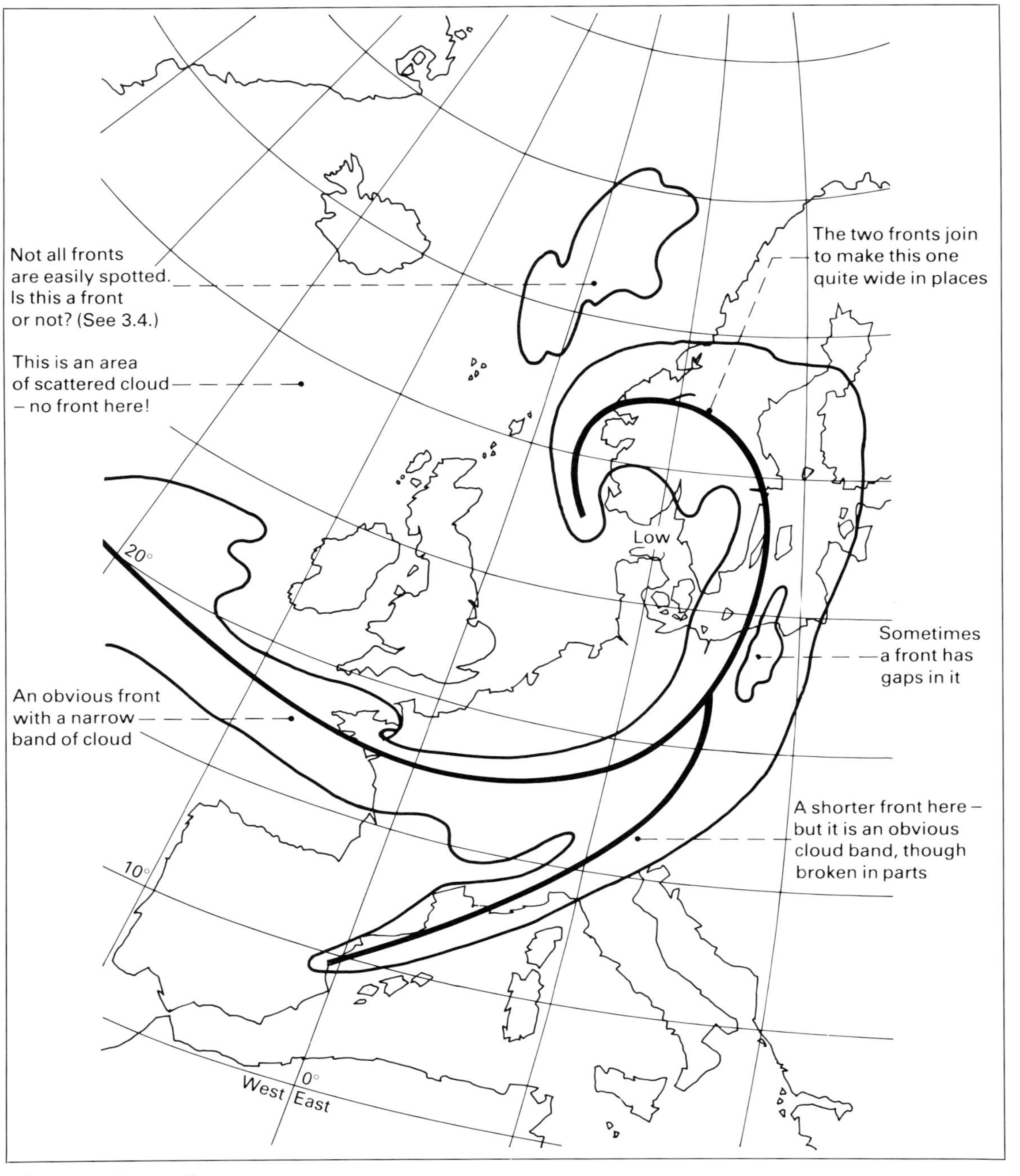

Fronts on Image 5

ACTIVITIES

1 Either use your own satellite image, together with a media weather map for the same day, or Image 9 and the weather map in Appendix 5, for this activity.

- a Look for a line of clouds that might make up a front line.
- b Mark these clouds on a copy of the right base map, and then mark on the front, using a black line.

  You can check your answer using information from a media weather map.

3.3

2 If you used your own satellite image for Activity 1 you can track the movement of the fronts. Look at another satellite image for later on the same day, and on your map mark the new position of the front or fronts. You should be able to work out the distance the depression has moved and the direction it has moved in, and also calculate the speed of the front's movement.

You can use Image 10 for this if you did Activity 1 using Image 9. Image 10 shows the depression at a later time.

3.4 Warm Fronts, Cold Fronts and Occluded Fronts

Most depressions are made up of two or three fronts – a *warm* front, a *cold* front and often an *occluded* front.

On a weather map, each front is marked with black lines with different symbols on them to show what kind of front it is. The side of the line the symbol is on shows the direction it is moving in. In Europe, most depressions usually move in from the Atlantic Ocean in the west.

The diagram below should help you to understand about the different kinds of front.

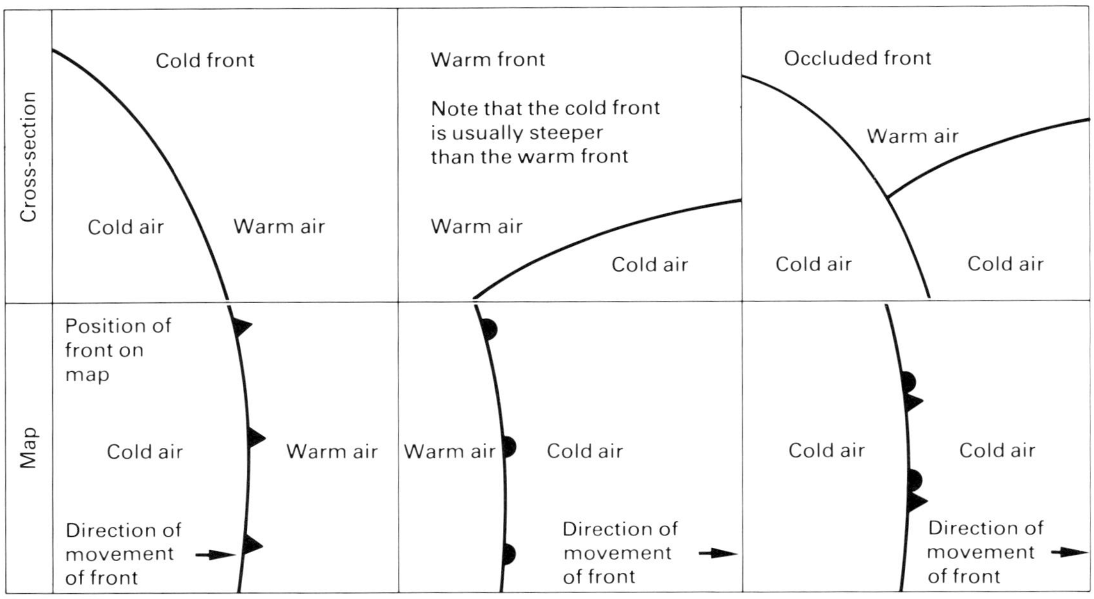

These fronts join together to form a typical depression pattern, shown in the diagram below.

Look at Image 5 together with the sketch on page 51. You should easily be able to see the three fronts that make up this depression on the satellite image.

3.4

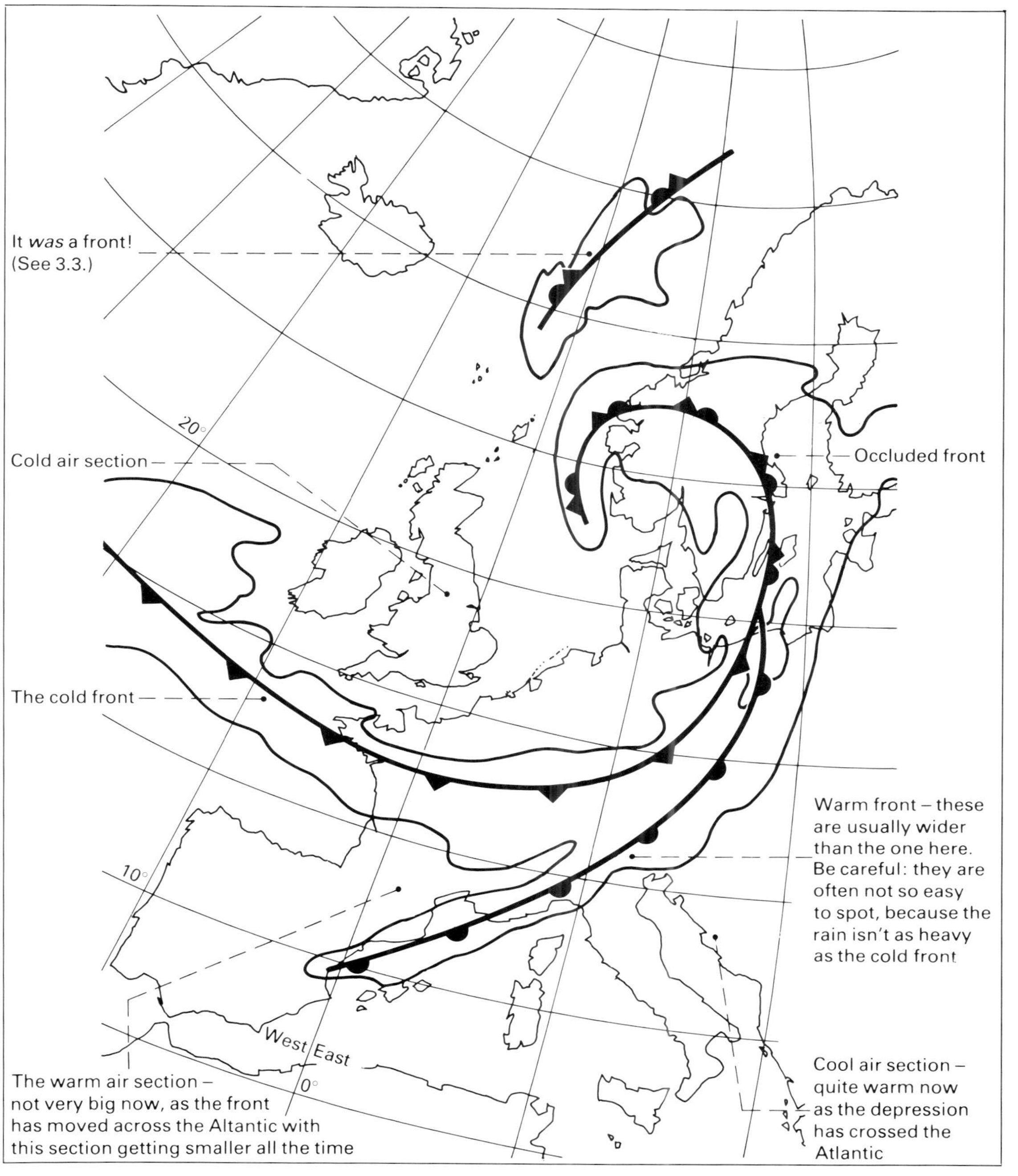

A depression on Image 5

ACTIVITIES

1 On a copy of the diagram and table on page 52, fill in the table carefully using the information in the diagram. When you have completed it, you should have a good summary of some of the characteristics of a depression which will help you to identify the different fronts on a satellite image.

3.4

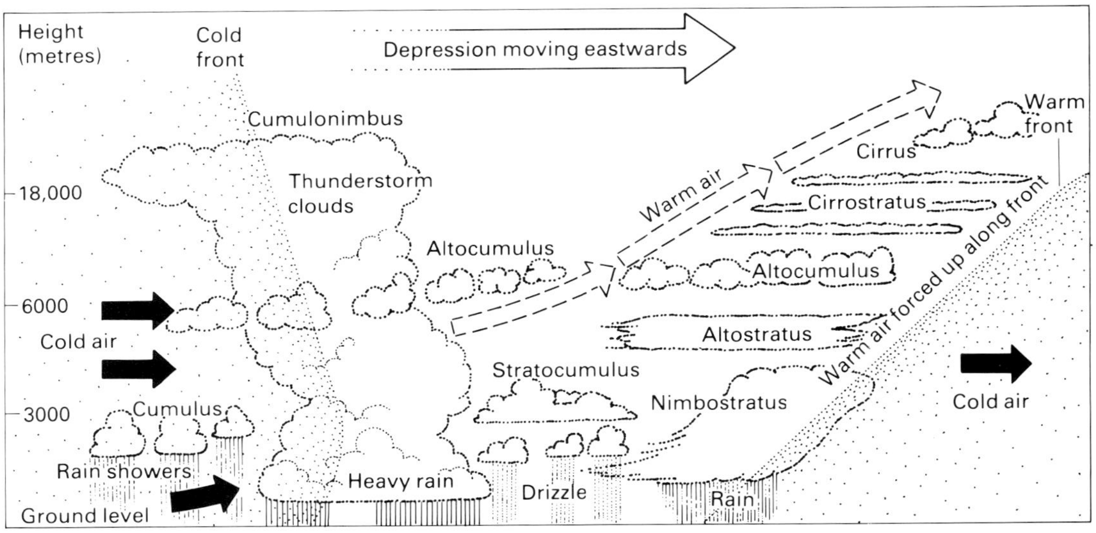

Characteristic	Warm front	Cold front*	Words to choose from
Angle of front			Gentle/steep
Height of clouds			Low to medium/high and towering
Length of time rain would fall			Long period/short period
Heaviness of rain			Heavy/quite gentle
Width of front on satellite image			Quite wide/quite narrow
Brightness of cloud on a V satellite image			Very bright and continuous/quite bright and sometimes less continuous
Brightness of cloud on an IR satellite image			Very bright and continuous/white-grey and sometimes less continuous

* The characteristics of an occluded front are similar to those of the cold front, although sometimes the occluded front is wider.

3.4

2 Either use your own satellite image, together with a media weather map for the same day, or Image 9 and the weather map in Appendix 5, for this activity.

Look at the satellite image of a depression, and on a copy of the right base map, mark on and label:

a the warm and cold fronts;
b areas where it might be raining; and
c areas of warm and cold air (warm and cold sectors).

3.5

The Weather in a Depression

The weather in a depression usually follows a set pattern. As depressions move over the UK, they often bring with them a predictable sequence of weather.

ACTIVITY

Because depressions are such large features and they are made of air masses that you can't actually see, they are often difficult to understand and imagine.

This activity involves making a simple model of a depression. It has been designed to help you imagine a depression in three dimensions. You should each have a copy on card of the depression model (Appendix 9).

You should firstly colour the different sectors of the depression to show the different air temperatures, and then fold and glue the model as shown in the diagram.

Then you should use all the information on the model to fill in a copy of the table below, which highlights some of the weather changes as a depression passes over.

Characteristic	**Place A**	**Place B**	**Place C**
Cloud height			
Cloud type			
Rainfall			
Temperature			
Pressure			
Wind speed			
Wind direction			

3.5

COURSEWORK ACTIVITY

This is a group activity because it involves making a lot of measurements of the weather over a short space of time.

You should set out to investigate this question: 'What is the weather like during the passing of a depression?'

Firstly, you should look out for a good depression coming towards the UK – you will need to spot it whilst it is in mid-Atlantic so as to be ready for it when it arrives! You can do this by listening to media weather forecasts or getting hold of a newspaper weather map.

Ideally you should choose a depression that will pass over the UK during the daytime, so that all the measurements can be made during school hours. It doesn't matter if you haven't the time or resources to measure the weather during the whole of a depression – measuring how the weather changes across just one of the fronts would be useful.

At regular intervals, such as once every hour, weather station measurements should be taken of the following things:

a temperature;
b rainfall – you will need an accurate measuring cylinder for this;
c wind speed and direction;
d cloud cover;
e cloud type;
f pressure; and
g humidity.

If possible, if you have a Meteosat system, you should get a print-out of a satellite image each hour. It would also be a good idea to take photographs of the weather at the same time as the weather measurements.

When you have all the results together, you should be able to make a large poster showing how the weather changes as a depression passes overhead. You could follow the poster design below.

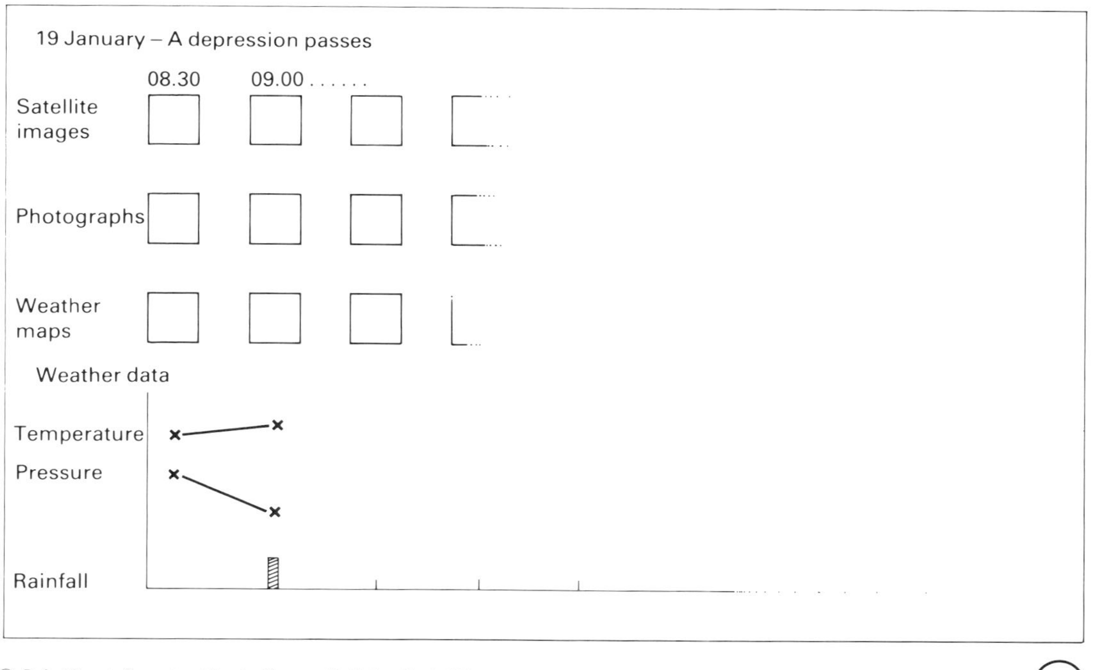

3.5

It might be possible after you have made the poster to contact the Meteorological Office to get hold of regular weather maps.

When you have made the poster, you could describe the passage of the depressions and also try to link together some of the weather characteristics. For example, what was the relationship between pressure and rainfall, or between temperature and cloud cover?

3.6 A Changing Depression

Depressions that affect the UK usually begin over the Atlantic Ocean and move from there towards the UK. Depressions that have passed over the UK move over continental Europe, where they die out.

It is often the case that the cold front is moving faster than the warm front, so the warm sector gradually gets smaller and smaller as the depression develops. Sometimes the cold front catches up with the warm front completely, and begins to lift the warm sector of air off the ground. There is still a front, but it is now separating two cold or cool air masses. The front is called an occluded front, and usually brings with it heavy rain and thick cloud. It is often the case that depressions are occluded before they pass over the UK.

The weather maps below show a depression moving across the Atlantic. The maps have been simplified to show only one depression – normally one depression after another comes across the Atlantic during the winter.

An occluded front is visible on Image 5.

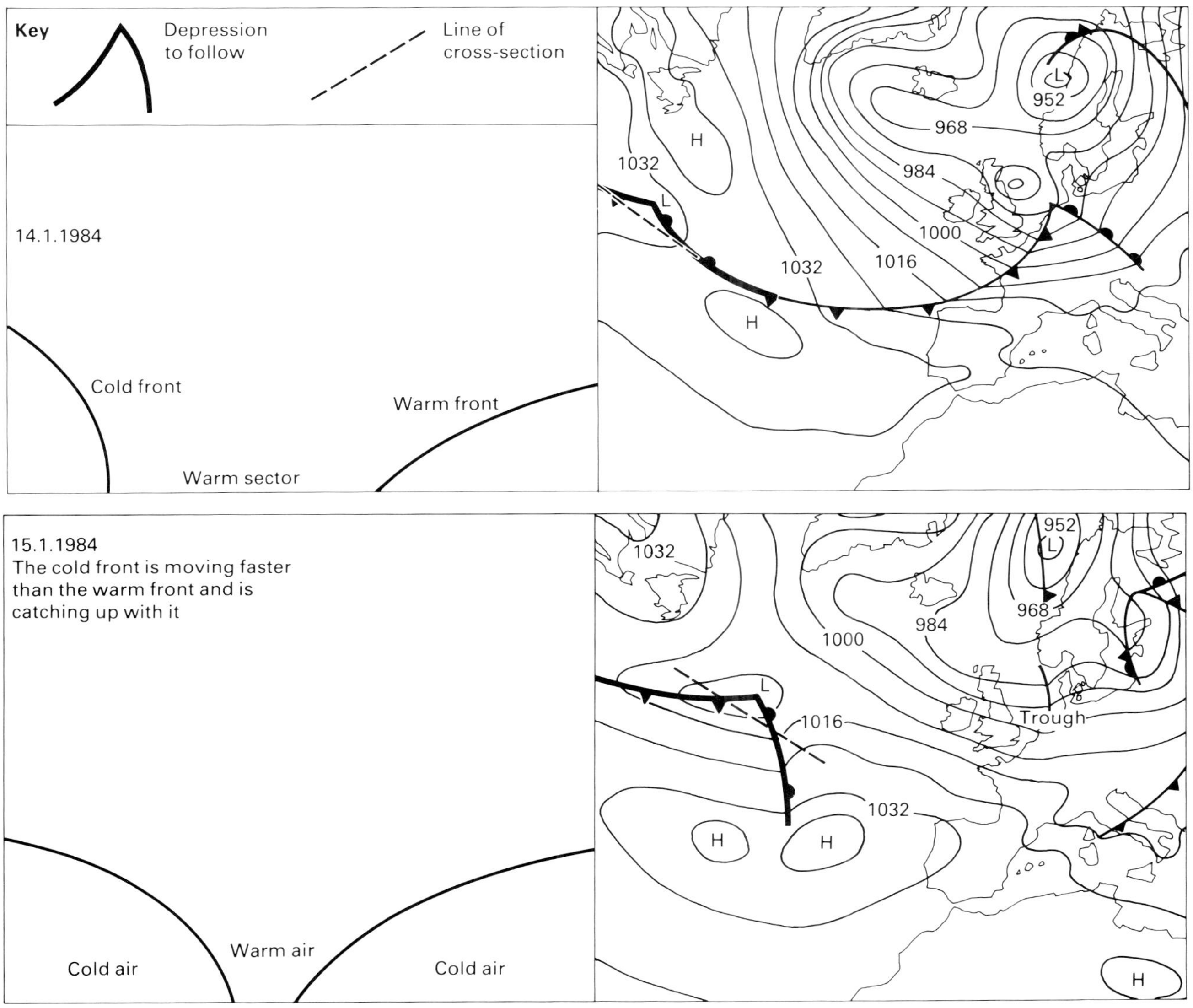

Cross-sections of a depression or 14–17 January 1984 (continues on page 58)

3.6

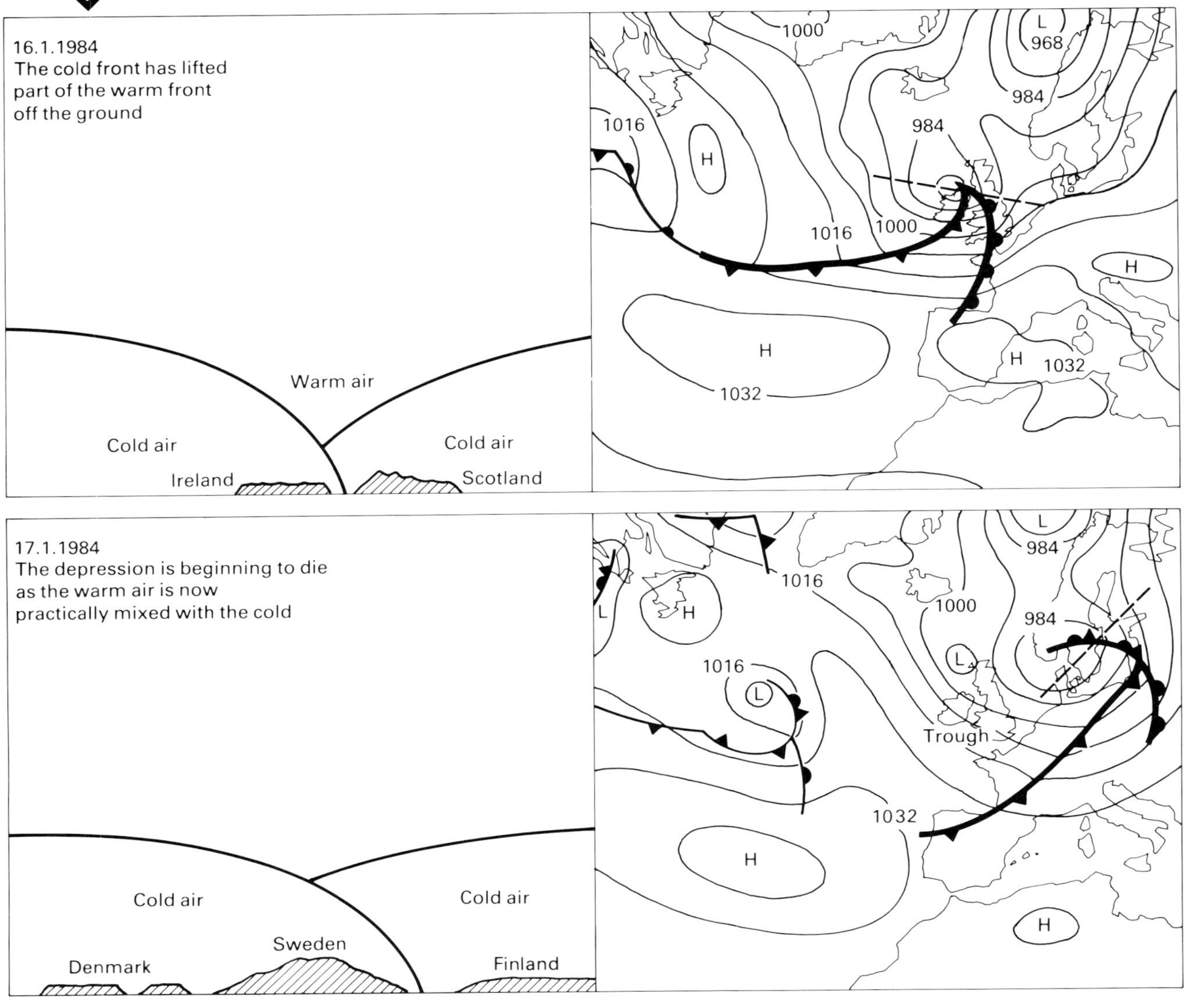

Cross-sections of a depression on 14–17 January 1984 (continued)

COURSEWORK ACTIVITY

You might like to study depressions in a bit more depth. To do this you could investigate this question: 'Do depressions follow a typical sequence of development?'

Firstly, you should look out for an approaching depression. It might help to listen to weather forecasts on radio and television to find out when one is on its way.

As soon as the depression appears on a satellite image, map the position of the depression as frequently as possible. If you have a Meteosat receiver you can do this at least once an hour. You should continue doing this until the depression disappears.

3.6

When you have mapped the movement of the depression you can make a number of calculations, such as the speed it moved; whether it slowed down or speeded up at any point; whether it moved along a smooth path or changed direction frequently; and so on.

If you are able to map the paths of a number of depressions you can see whether they fit in with the typical depression tracks shown on the maps below.

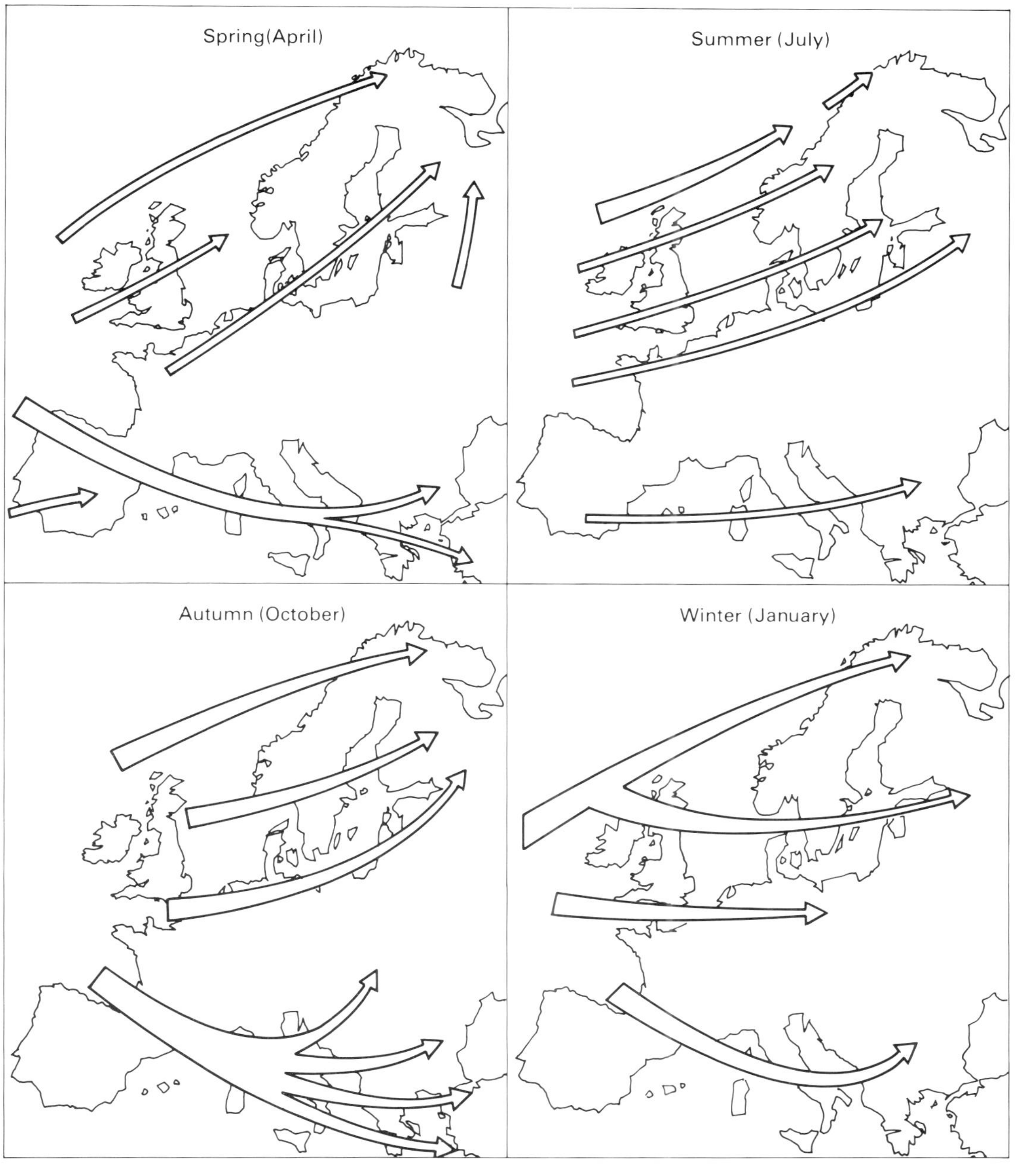

The main paths followed by depressions over western Europe

3.7

Anticyclones

Anticyclones give us totally different weather to depressions. You have already seen that anticyclones are areas of high pressure that tend to bring good weather with them.

Look at Image 6 alongside the sketch below. There is high pressure over central Europe, and the UK is just on the edge of it. Most of the UK has clear weather.

Anticyclones differ from depressions in another way too. They are usually quite stable features and can stay over a particular location for days and sometimes weeks. The droughts we have had in the UK in the past 10 years have been because an anticyclone was 'stuck' over the UK for a long time.

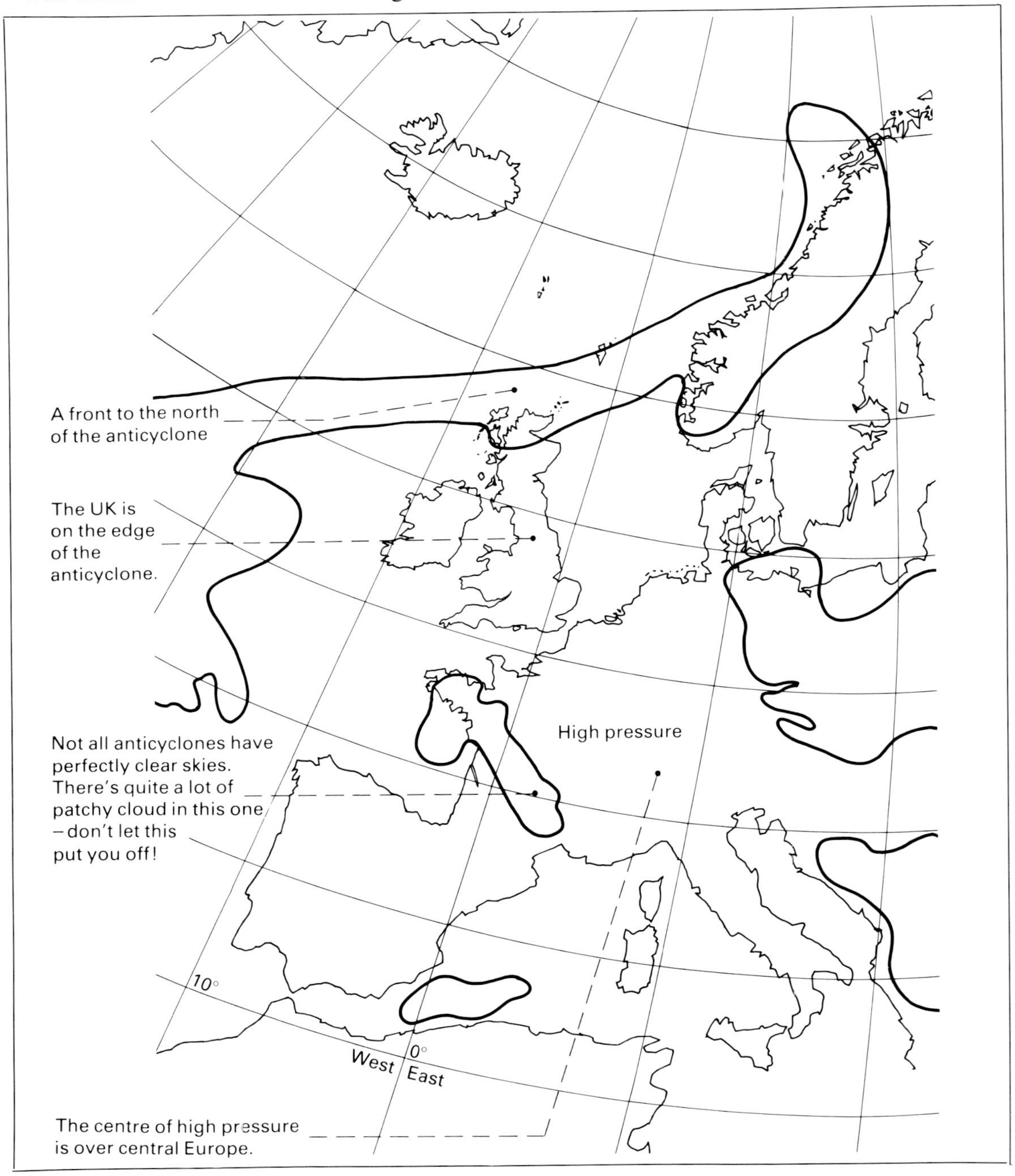

An anticyclone on Image 6

3.7 Anticyclones 'block' the way of depressions sometimes. They divert them either to the north or to the south of the UK.

There are big differences between anticyclones in the winter and in the summer.

In summer, the fine weather and clear atmosphere allow the sun to shine for long periods. The ground is heated and this in turn heats the air. An anticyclone in summer will show up on a satellite image as a large area where the land is clearly seen. There may be some high cloud giving a hazy effect.

In winter, it isn't warm because the sun isn't high enough in the sky and the clear skies mean that the little heat the earth does have can escape. Temperatures are therefore cold. Anticyclones still give clear conditions in winter but thin cloud is common, giving overcast but dry conditions.

Fog or mist also often happens with an anticyclone. In summer this is quickly evaporated, but in winter it can last all day. Thick fog or mist will not show up on a visible image, but it will on an infra-red image.

One particular kind of fog is called *advection fog*. The diagram below shows one way this forms. This kind of fog shows up easily on satellite images. Look at the area over Denmark on Image 1 – this is advection fog caused largely by the cooling of moist air over the North Sea.

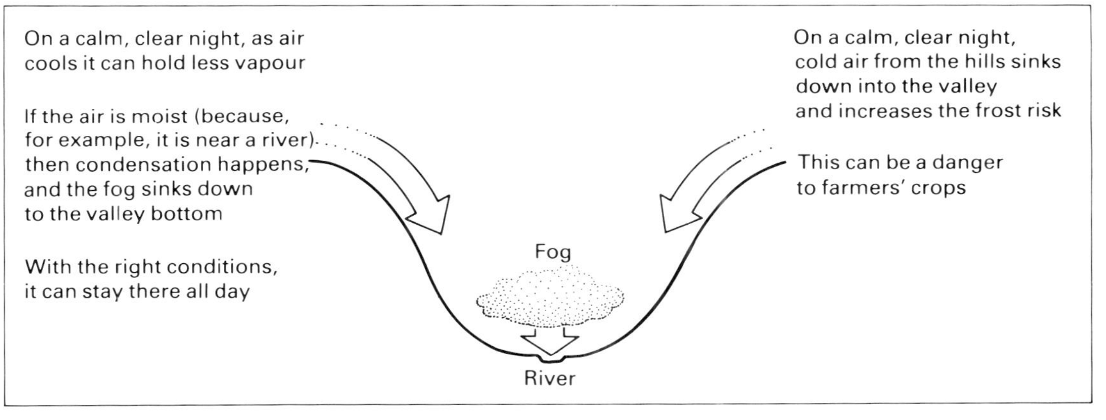

Two valley problems

ACTIVITIES

1 Look through all the satellite images included in this pack and try to work out which ones show anticyclones.

2 Advection fog can be a menace, especially to transport networks such as air routes and motorways.

Look at the map on page 62. From a meteorologist's point of view, which is the best site for the city airport, and which is the best route for the new motorway between the two cities? Explain your choices.

3.7

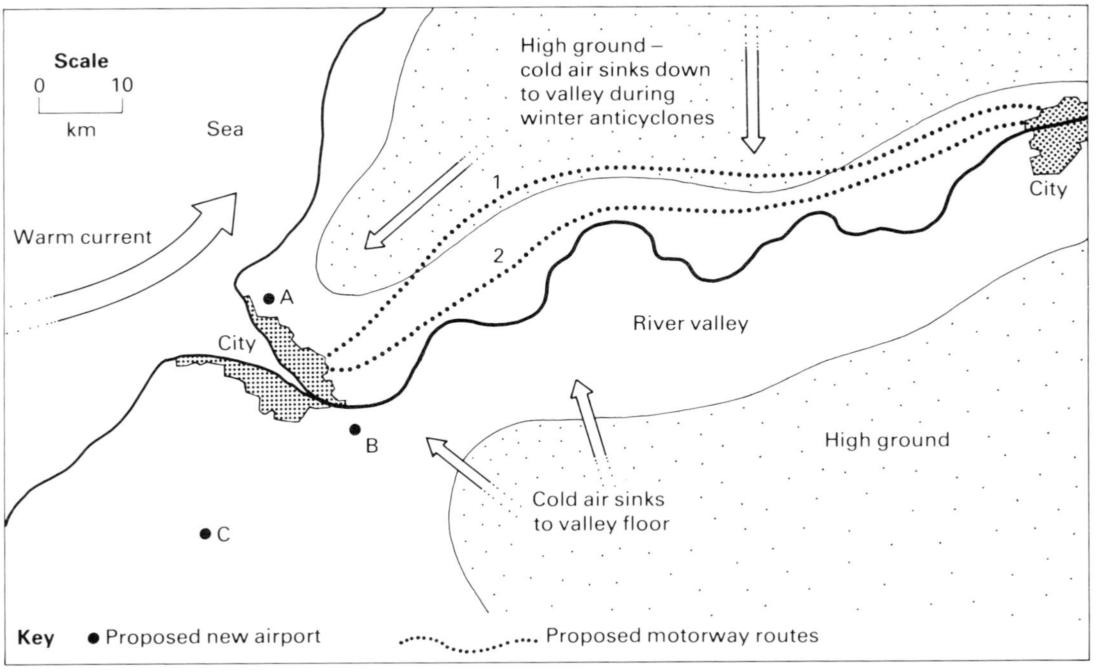

COURSEWORK

ACTIVITY

You could investigate this question: 'What is the relationship between pressure, clouds and rainfall?'

You can either use Images 3, 4 and 6 together with the weather maps you will find in Appendix 5, or use your own images together with media weather maps.

If you are using the images in the pack, you should choose a number of weather stations in Europe, and from each of the satellite images work out how much cloud cover each one has. You won't be able to be very detailed – you will be able to say that a place is cloudy, has scattered cloud or has no cloud. Then you should find out the pressure at each of your chosen weather stations, and try to spot any relationship between the two things.

If you are using your own images, you can either carry out a similar activity, or keep a pressure and cloud cover diary for your weather station. If you keep the diary for a month or so, you should be able to see some relationship between pressure and cloud cover.

For both investigations you should present your data using a scatter graph, with the cloud cover along one axis, and pressure along the other.

OBJECTIVES

CHECKLIST

3.1

Depressions and Anticyclones

Do you

Tick here

- know and understand what depressions and anticyclones are? ☐
- know and understand how winds are formed? ☐

Can you

- identify areas of high and low pressure on a satellite image? ☐
- draw an isobar map? ☐
- mark wind strength and direction on an isobar map? ☐

3.2

Depressions and Air Masses

Do you

- know and understand about air masses and their different characteristics? ☐
- understand some of the problems very cold weather can cause old people? ☐

Can you

- find a depression on a satellite image? ☐
- find different air masses on a satellite image? ☐
- work out what the wind will be like that will affect the UK? ☐
- use newspaper weather data (or data from another media source)? ☐

3.3

Fronts

Do you

- know and understand what a front is? ☐

Can you

- find a front on a satellite image? ☐
- work out the speed of movement of a front? ☐

3.4

Warm Fronts, Cold Fronts and Occluded Fronts

Do you

- know and understand about warm and cold fronts? ☐

Can you

- make up a summary table using information in a diagram? ☐
- find warm and cold fronts on a satellite image? ☐

3.5 The Weather in a Depression

Do you

- know what weather to expect as a depression passes overhead? ☐

Can you

- make up a model of a depression? ☐
- get information from a model to fill in a table? ☐
- make frequent recordings of weather during a depression? ☐
- take photographs of weather? ☐
- plot weather data in an accurate and interesting way? ☐
- make an attractive and informative poster? ☐

3.6 A Changing Depression

Do you

- know and understand what an occluded front is? ☐
- know how a depression develops over time? ☐

Can you

- identify an occluded front on a satellite image? ☐
- trace the development of a depression through a sequence of satellite images, and make a weather map at each stage? ☐

3.7 Anticyclones

Do you

- know and understand about low pressure, high pressure and anticyclones? ☐

Can you

- identify an anticyclone on a weather satellite image? ☐
- use weather information to decide on the best routes and locations for transport facilities? ☐
- read pressure from a weather map? ☐
- correlate (link) information from two maps? ☐
- draw a scatter graph from the information? ☐

4.1

High and Low Ground

The higher the land, the colder it is. Temperatures usually drop by about 6 °C for every 1000 m of altitude.

An infra-red image might show some of these variations in temperature which are due to the height of the land. This is only possible, of course, when there isn't much cloud cover and you can see the land directly. As the higher land is cooler, it emits less infra-red radiation and so appears as a light grey shade on the satellite image.

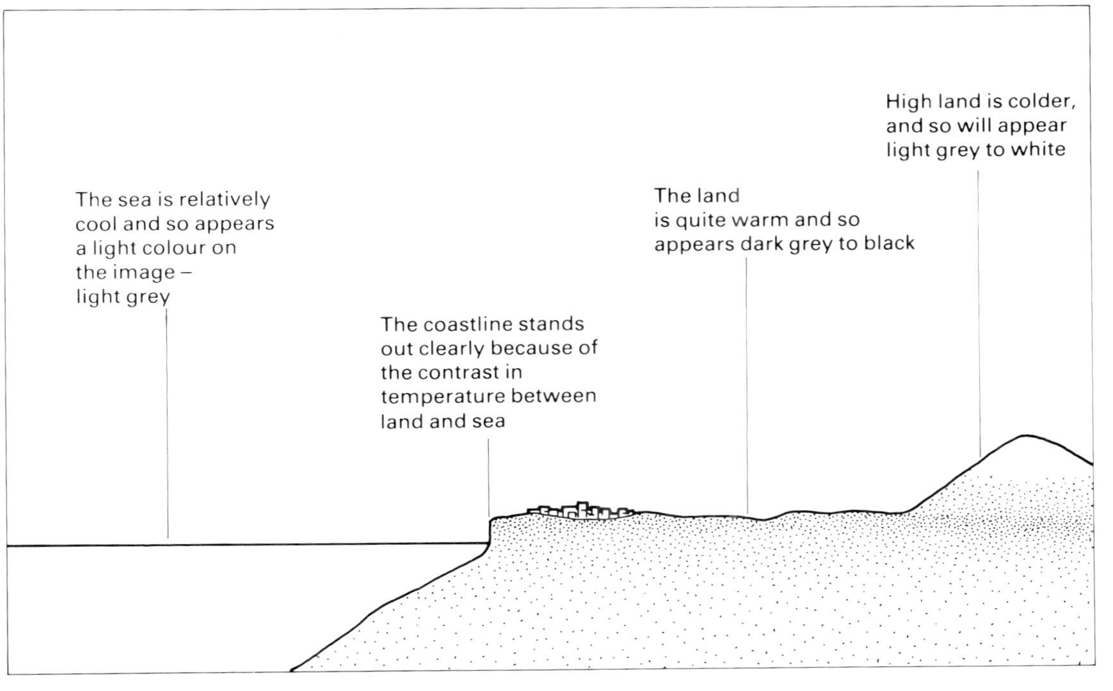

A daytime infra-red image in the summer

Look at Image 7 alongside the map on page 66. Some upland areas are clearly visible on this image. You can see the higher land down the centre of Italy, the higher land of Norway and the mountains between France and Spain quite clearly. Do be careful, however – it is easy sometimes to mistake clouds for high land!

4.1

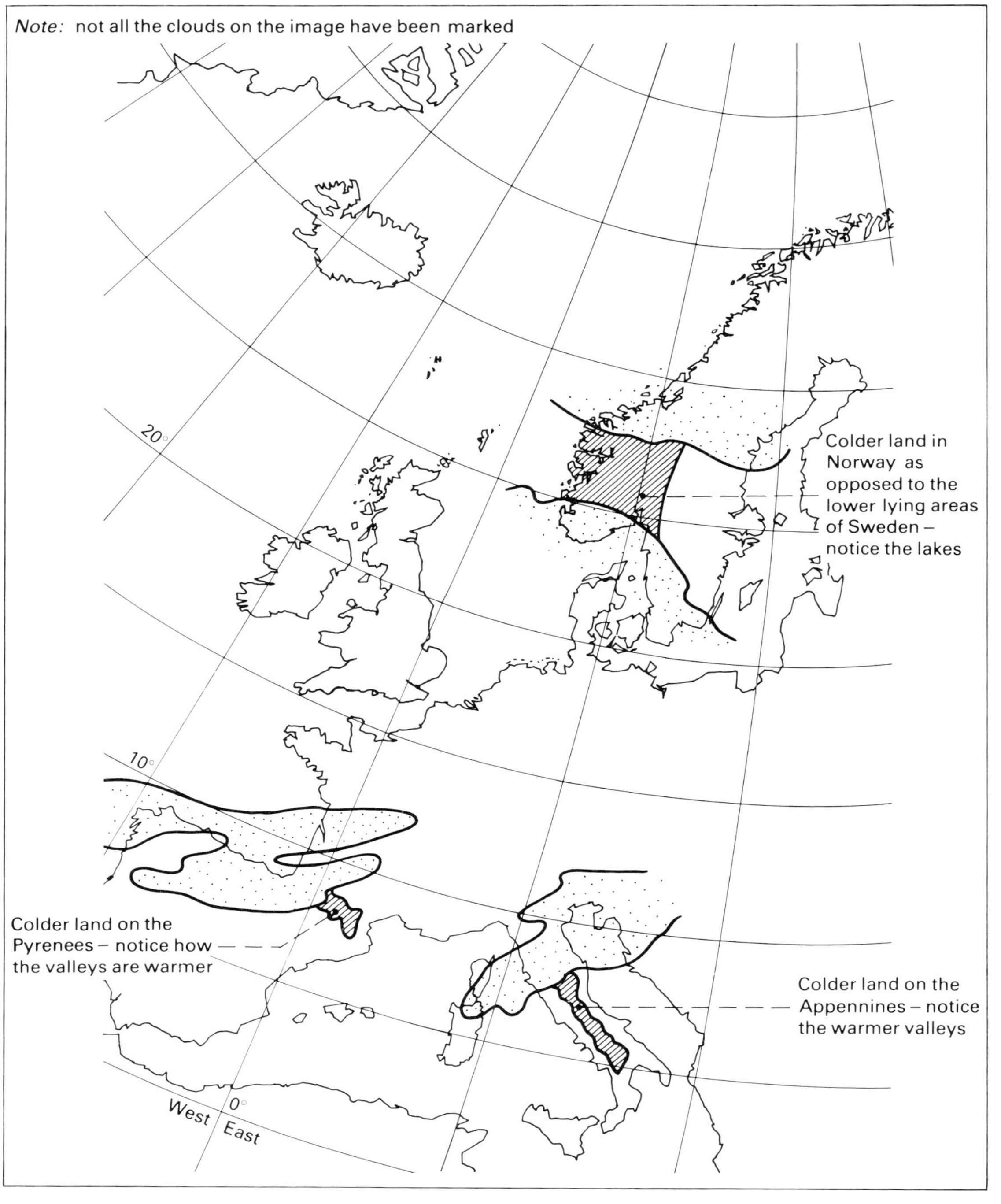

High and low ground on Image 7

COURSEWORK ACTIVITIES

1 Using a visible image, find a large area of Europe that is not covered by cloud. Use an infra-red image to find the coldest parts of the same area, and shade them on a copy of the right base map.

Now find a map in your atlas which shows the height of the land, and on another copy of the base map shade the highest areas of land and name them.

4.1

Finally, fill in a copy of the table below, and say whether the statement that higher areas of land show up colder on infra-red satellite images is true or not.

You might also like to try using a visible image for the same activity. High land is often marked on visible images by the presence of clouds 'sitting' over the tops of the mountains. This is because air rises up a mountain side, and as it does so it cools and the water vapour may condense out to form clouds.

Country or area under study: ....

Date: .... **Time of image:** ....

Name of area of high land	Shade on satellite image

2 You can link the height of a location with temperature in different ways as well.

For example, you can compare temperature and relief maps in your atlas. Or, if you look at the climate graphs in your atlas, you can note the mean temperature for a number of locations, together with their heights above sea level, and draw a scatter graph to find the relationship between height and temperature.

COURSEWORK ACTIVITY

An interesting coursework activity would be to look at the temperature changes across a valley. You could link this sort of study to aspects of slopes and land use.

4.2

North and South

Average temperatures get lower as you move away from the equator towards the poles, both north and south. Look at a temperature map for the world in your atlas, and you will see that a north–south pattern applies to Europe as a whole, as well as to some individual countries which have a large range of latitude.

This temperature variation happens because the angle of the sun is lower the further north or south of the equator you go. This diagram explains why.

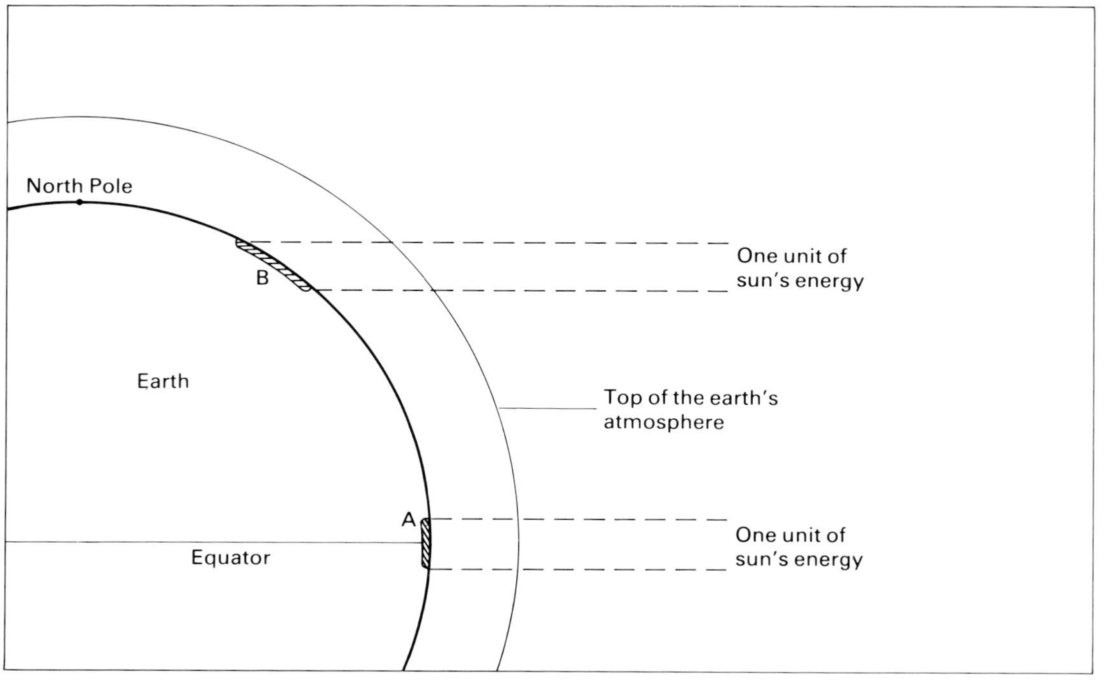

At areas A and B, one unit of energy from the sun has passed through the earth's atmosphere to heat the ground. Area A, at the equator, will be warmer than area B, which is about where the UK is, for a number of reasons:

1 At B, the one unit of energy has a greater ground area to heat compared with A, so the heat is more spread out.

2 At B, the one unit of energy has more atmosphere to pass through, so more energy will be scattered or absorbed before it reaches the ground.

Look at Image 8 alongside the sketch on page 69, and you can see that the infra-red image shows how temperatures can decline from south to north.

There is a complicating factor, though, which will be looked at in 4.3 – the seas around Europe affect the temperatures of the coastal areas.

4.2

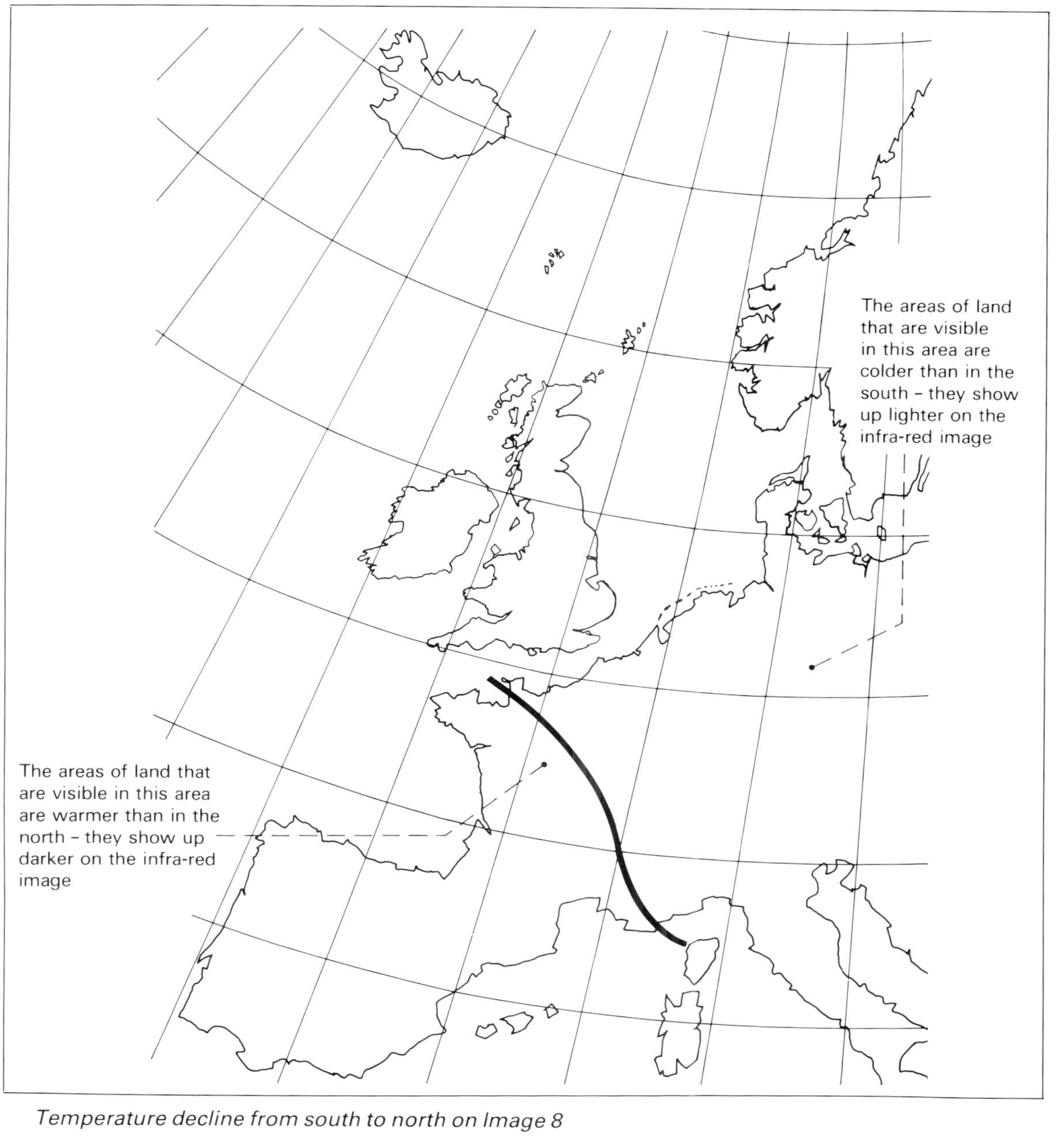

Temperature decline from south to north on Image 8

ACTIVITIES

1 Using a visible image, find a large area of Europe that is not covered by cloud. The area you choose should show quite a large north–south distance.

Now use an infra-red image of the same area to try to discover any patterns of temperature difference. Shade this pattern on a copy of the right base map, then shade the map with a line or colour shading.

4.2

Describe the pattern of temperature variations. Is there a gradual cooling as you go from south to north on the image?

2 Find a table of temperatures for the area from the next day's newspaper and mark the actual temperatures of some places on your map. Is the pattern of temperature change similar to the one you have found?

4.3 Distance from the Sea

In western Europe, distance from the sea will affect the temperature experienced at any place.

In winter, the sea is warmer than the land. This is because the North Atlantic drift, a warm ocean current which comes from the Caribbean across the Atlantic Ocean, keeps western coastal areas warmer than they should be for their latitude. Air which travels over this 'warm' ocean is also warmed. In winter, therefore, the coastal areas of western Europe are warmer than continental ones, because the North Atlantic drift acts on them like a hot-water bottle.

In summer, the pattern is the exact opposite – the sea is now cooler than the land. Areas of land warm up very quickly in summer, but the sea takes longer. The result is that inland areas tend to be a lot warmer than places along the coast.

The map below shows the kind of temperature pattern you might expect to see in the winter across Europe, as the sea keeps coastal areas warmer than ones inland.

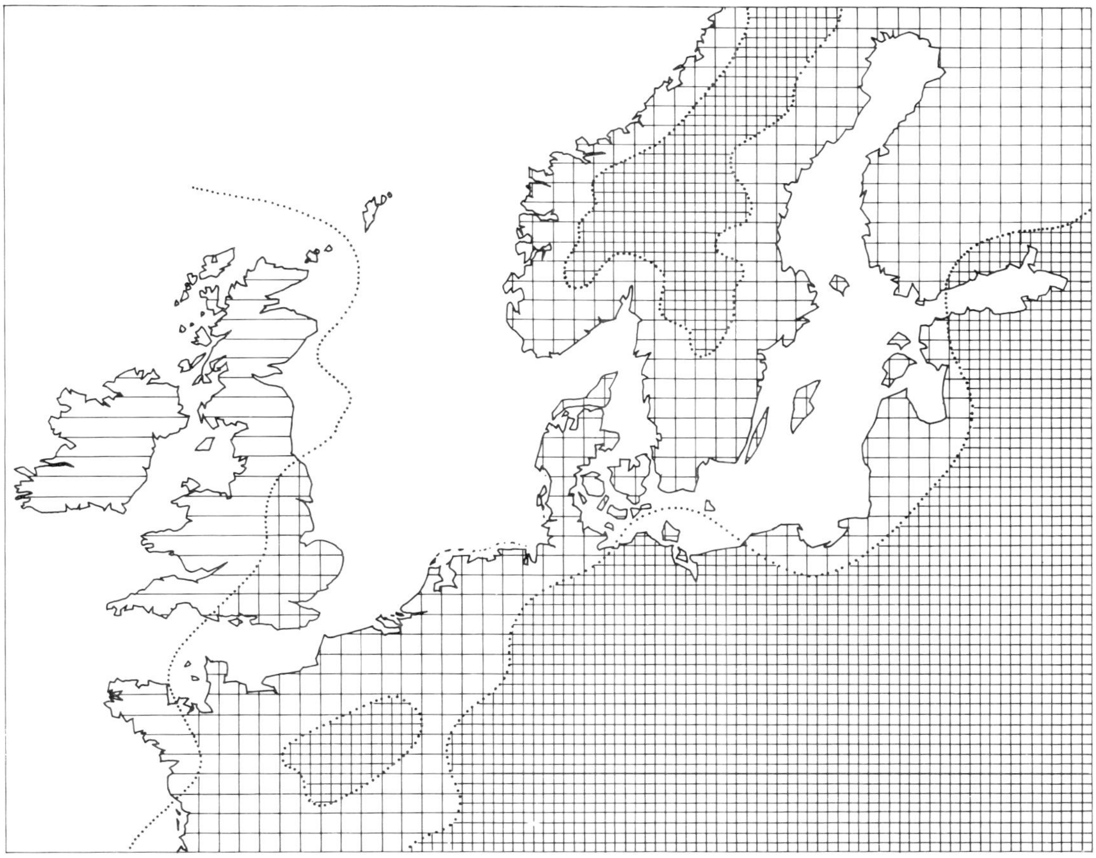

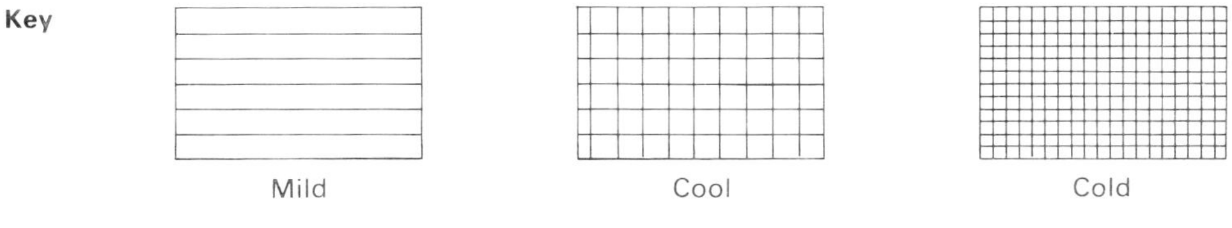

4.3

ACTIVITIES

1 Using a visible image, find a large area of Europe that is not covered by cloud. The image you choose should show both coastal and inland areas. You will get better results if you try this activity during the winter.

Now use an infra-red image of the same area to try to discovery any patterns of temperature difference. Shade this pattern on a copy of the right base map, using a similar shading to the one on page 71, or a colour shading.

Describe the pattern of temperature variations. Is there a general cooling as you go from coastal to inland areas?

2 Choose from newspapers a number of named weather stations fairly close to a cross-section line which runs from a west coast area, such as Ireland, through to as far east as your image will let you go.

The line doesn't have to be straight, but it would help if the places were about the same height above sea-level. A good list would include places like Shannon, Paris, Hanover, Berlin and Warsaw. There are plenty of other places to choose from, as long as their temperatures are available in a daily newspaper. As these temperatures are recorded at midday, it would be useful to look at an afternoon infra-red image.

Plot the temperature figures as a line graph like the one below, and then add information from the satellite image in the form of bars to represent the shades on the image.

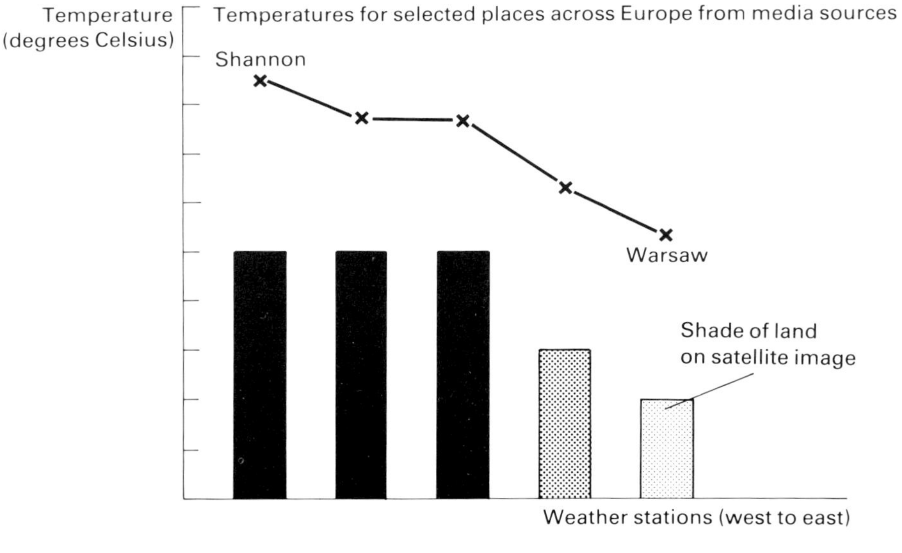

4.4 Urban Heat Islands

Built-up areas such as towns and cities tend to be warmer than the countryside areas which surround them. One researcher found out that temperatures across London one winter's night were 8 °C higher in the centre, compared with the suburban and rural locations.

The map below shows the results of this research. You can clearly see the temperature differences.

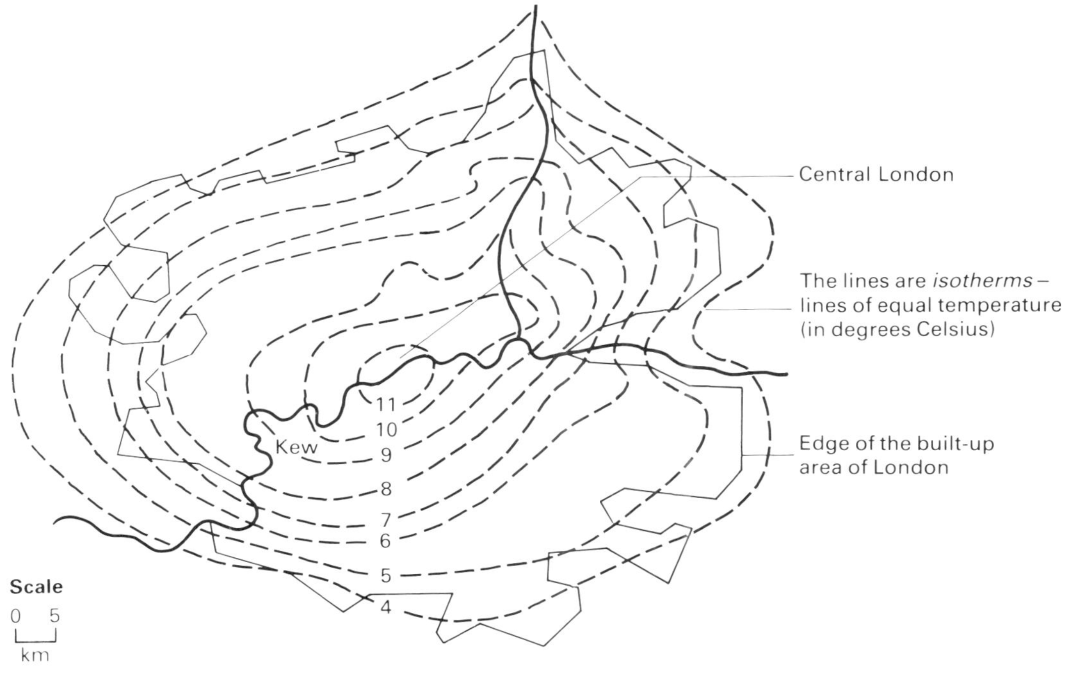

Because the pattern of isotherms (lines of equal temperature) looks like a contour pattern of an island, this difference in urban and rural temperatures is often called an *urban heat island*.

Urban heat islands happen because a number of things make cities warmer. Heat is given off by traffic and buildings; buildings, concrete and tarmac all absorb heat easily; radiated heat is reflected off buildings down to the ground again; and the urban atmosphere is dustier, and so reflects heat.

It is sometimes possible to pick out urban heat islands on infra-red satellite images. The best images to use are those taken in the night-time, as then the contrast between urban and rural areas is likely to be the greatest.

Urban areas are also sometimes to be seen on clear visible images, because large groups of buildings tend to reflect less light than countryside areas – especially in the summer. You should be able to pick out darker areas on visible images which mark the location of major urban areas.

Look at Image 1 and see if you can pick out London and Paris. Be careful, though: the darker areas on land don't always represent cities.

4.4

ACTIVITIES

1 Use an atlas to find the major urban areas of Europe and mark them on a copy of a suitable base map.

Compare your map of built-up areas with an infra-red image for the same area. Do the two maps match up in any way? Do the urban areas stand out as being warmer than surrounding rural ones?

2 Look at the map below, which shows the pattern of mean temperatures in Paris.

On a copy of this map draw the isotherms (lines of equal temperature) for 11 °C and 12 °C. The 11 °C isotherm has been started for you.

Describe the pattern the isotherms show. Which parts of the Paris area are the warmest and which parts the coolest?

What effect do you think that the urban heat island would have on the weather of Paris?

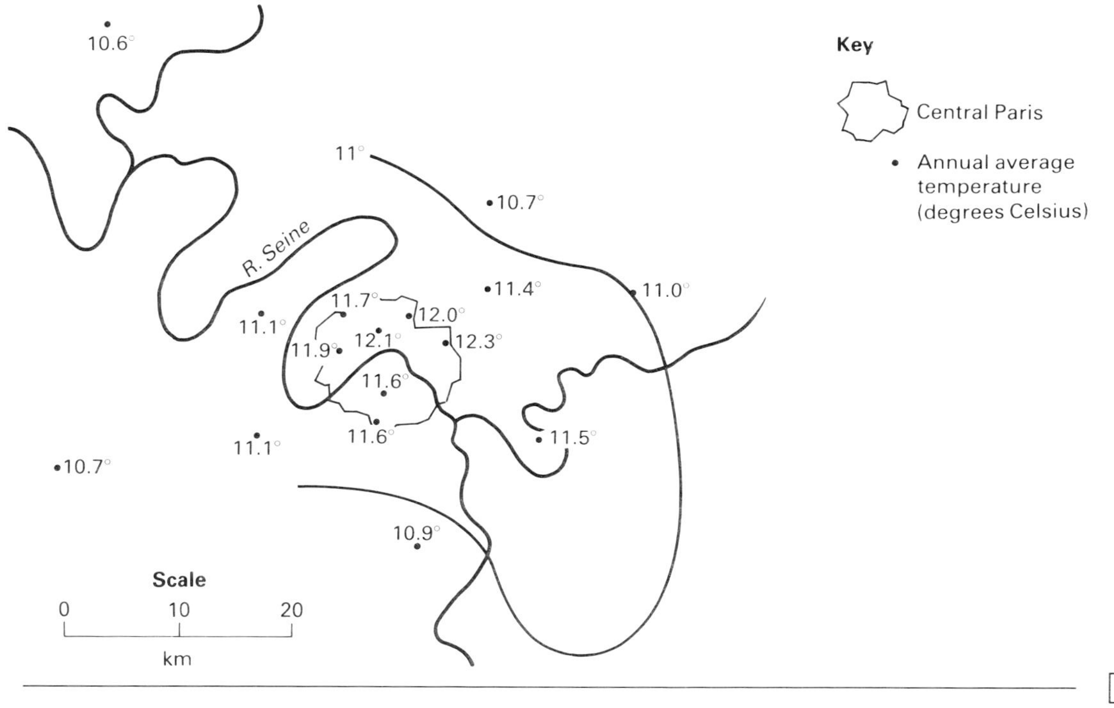

COURSEWORK

ACTIVITY

Test the idea that temperatures in a town are higher than in the surrounding countryside by doing the following fieldwork as a group.

To test this idea you should place a number of maximum and minimum thermometers along a transect going from the centre of a town to the outskirts.

4.4

This is more difficult than it seems. You have to choose the sites for your thermometers carefully, so that:

a they are safe from vandalism, and safe to visit to record the measurements.
b they are all in a similar position – they should all be the same height above the ground, for example, and should not be put in places where the results would be inaccurate.
c they can all be read at about the same time each day – unless you are just interested in the maximum and minimum temperatures.

You will also have to make other decisions in your testing of the idea, such as how long you are going to collect the results for and how to present your data when it has been collected.

4.5 Day and Night, Summer and Winter

Temperatures over Europe change over periods of time. This happens both in the short term – for example, between day and night – and also in the long term – for example, between summer and winter.

These changes can be seen on an infra-red satellite image very clearly, especially if you compare the relative temperatures of the land and sea. The diagrams below show what happens in just one weather situation.

If you look at Images 5 and 8, you will see they illustrate the contrasts between the summer and winter. They are summarised on the sketch on page 77.

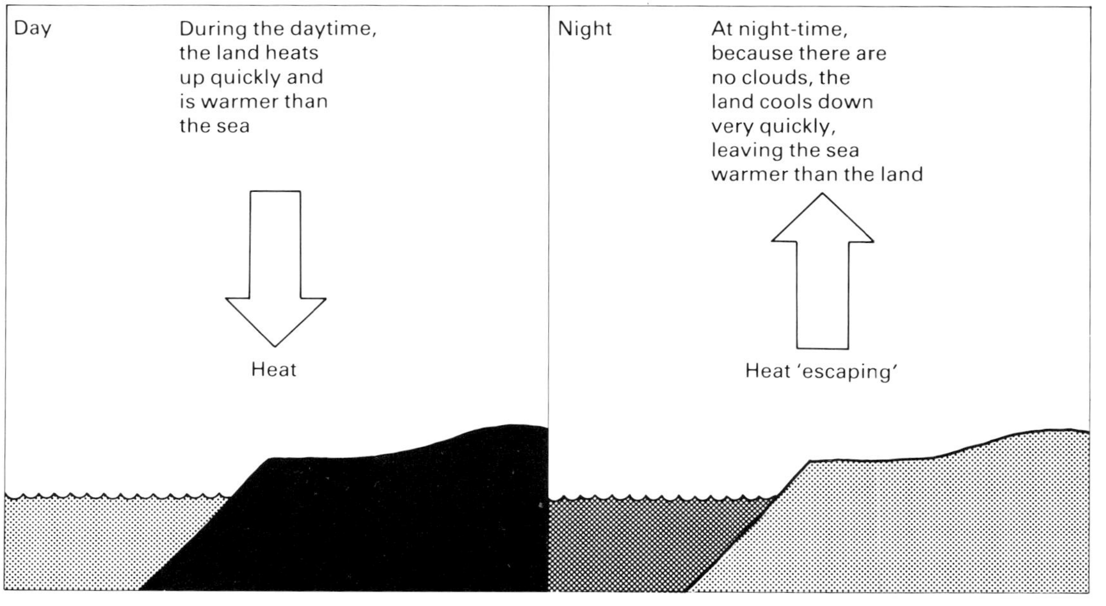

Temperature changes between day and night during a winter anticyclone

ACTIVITIES

1 You should obtain one daytime and one night-time infra-red satellite image for approximately the same area.

How has the temperature of the land changed between day and night? How has the relative temperature of land and sea changed between day and night?

2 If your school has the facility, it would be useful to build up a library of infra-red images to show how the temperatures change both between day and night, and during the year. These could be displayed in the form of a poster along with some appropriate diagrams, explaining why the changes take place.

4.5

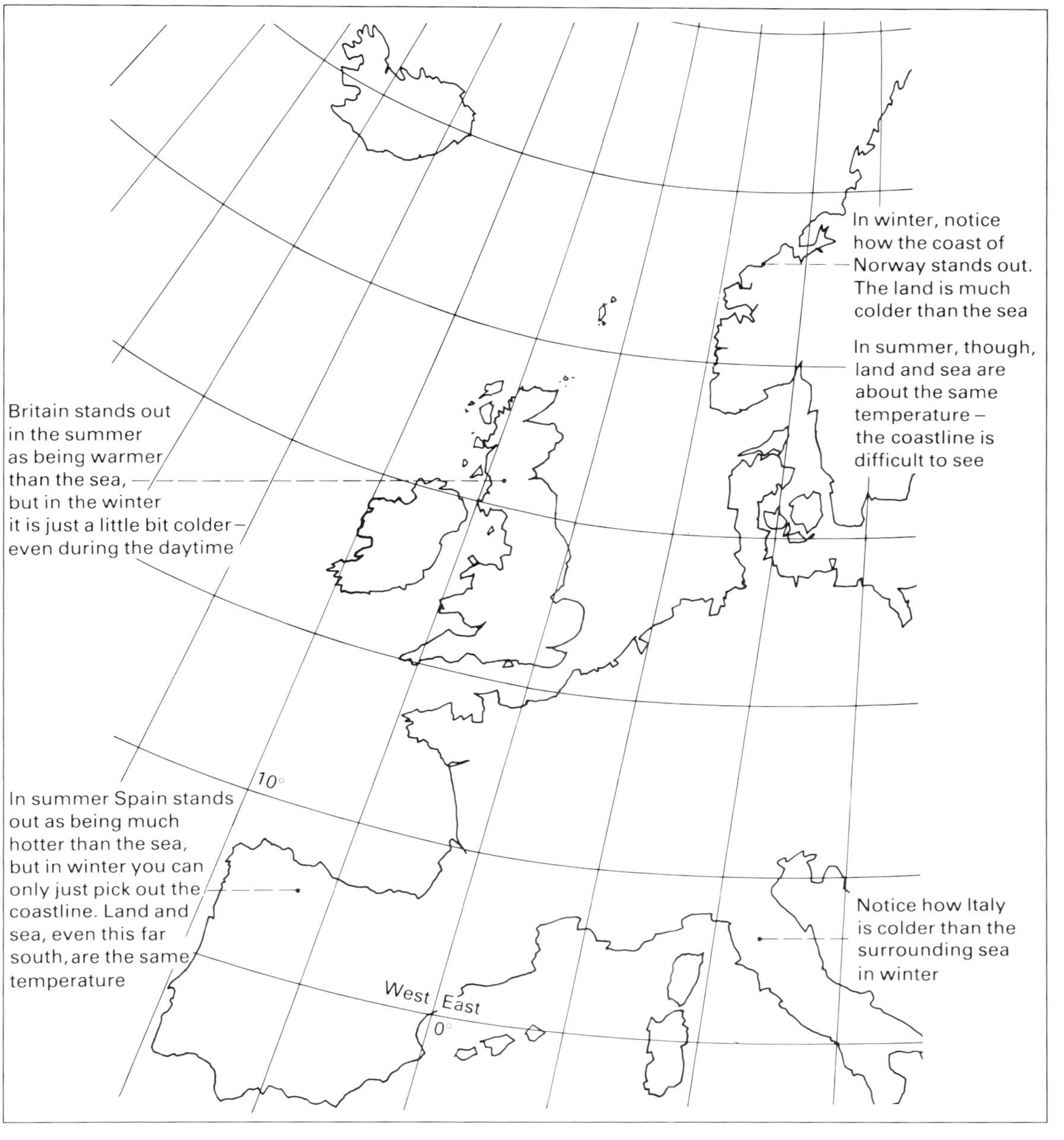

Temperature changes between Images 5 and 8

COURSEWORK

ACTIVITY

Test the idea that clouds will affect the maximum and minimum temperatures a place has.

The diagrams above and on page 76 show that cloud cover is likely to affect both maximum and minimum temperatures in different ways. For example, a cloudless night in summer will be colder than a night with a lot of cloud.

4.5

To test this idea, you should fill in a copy of the table below for a period of at least a week. The maximum and minimum temperatures can be measured in the usual way, and cloud cover during the day measured in *oktas* (eighths of cloud cover) by simply looking. The cloud cover during the night-time should be estimated using an infra-red satellite image.

Date	**Maximum temperature**	**Minimum temperature**	**Day cloud**	**Night cloud**

Once you have obtained your results you should represent your data in the form of graphs and diagrams, and then try to make links between the temperatures and cloud cover.

4.6

Ice on the Roads

One weather problem which affects large numbers of people in Britain in the winter is the icing of roads.

Ice on the roads can both cause accidents and generally slow down the traffic. In addition to the tragic loss of life accidents can sometimes involve, they cost a local authority money when the rescue services have to be called out to help. Ice also costs money if the traffic has to slow down too much, as freight journeys will take longer and therefore cost haulage companies more.

It is, however, possible to deal with the ice problem by spreading salt and grit on the roads. This is one of the many responsibilities of the Highways Authority. But stopping ice isn't simply a case of sending out the salting and gritting lorries every night through the winter just in case. Too much salt on the roads is a nuisance when it's not needed – it causes vehicles to rust more easily, and can also affect the structure of bridges. If it doesn't freeze, then a lot of money and effort will have been wasted.

On the other hand, if it does freeze when the Highways Authority hasn't salted the roads, then the result could be chaos. Obviously, therefore, each night during the winter the Authority will have a tricky decision to make – to send out the salting teams or not.

In parts of Wales this decision has been made easier, as in some areas special sensors have been placed at selected roadside locations to monitor road temperatures and the likelihood of freezing. The data from these sensors is fed into a computer in Swansea, which then issues ice warnings to each of the counties when necessary. The system is called ICELERT.

ACTIVITY

Imagine that you have been placed in charge of two of Dyfed's highways areas – St Clears and Carmarthen. It is your job to decide on the basis of a number of data reports whether or not to salt the roads you are responsible for. If you decide to salt the roads, and freezing does occur, then you earn yourself bonus points. On the other hand, if you make the wrong decision and either waste salt or are too mean with it, then points are taken away.

Each Highways Authority should have a copy of the Dyfed road map and the summary sheet, and a pack of five ICELERT cards.

The first thing each group should do is look carefully at the road map showing the routes you are responsible for. You are only interested in the A and B roads, and of course the M4. Although you would like to make all the roads safe when icing is likely to happen, this is just not possible in cost terms.

4.6

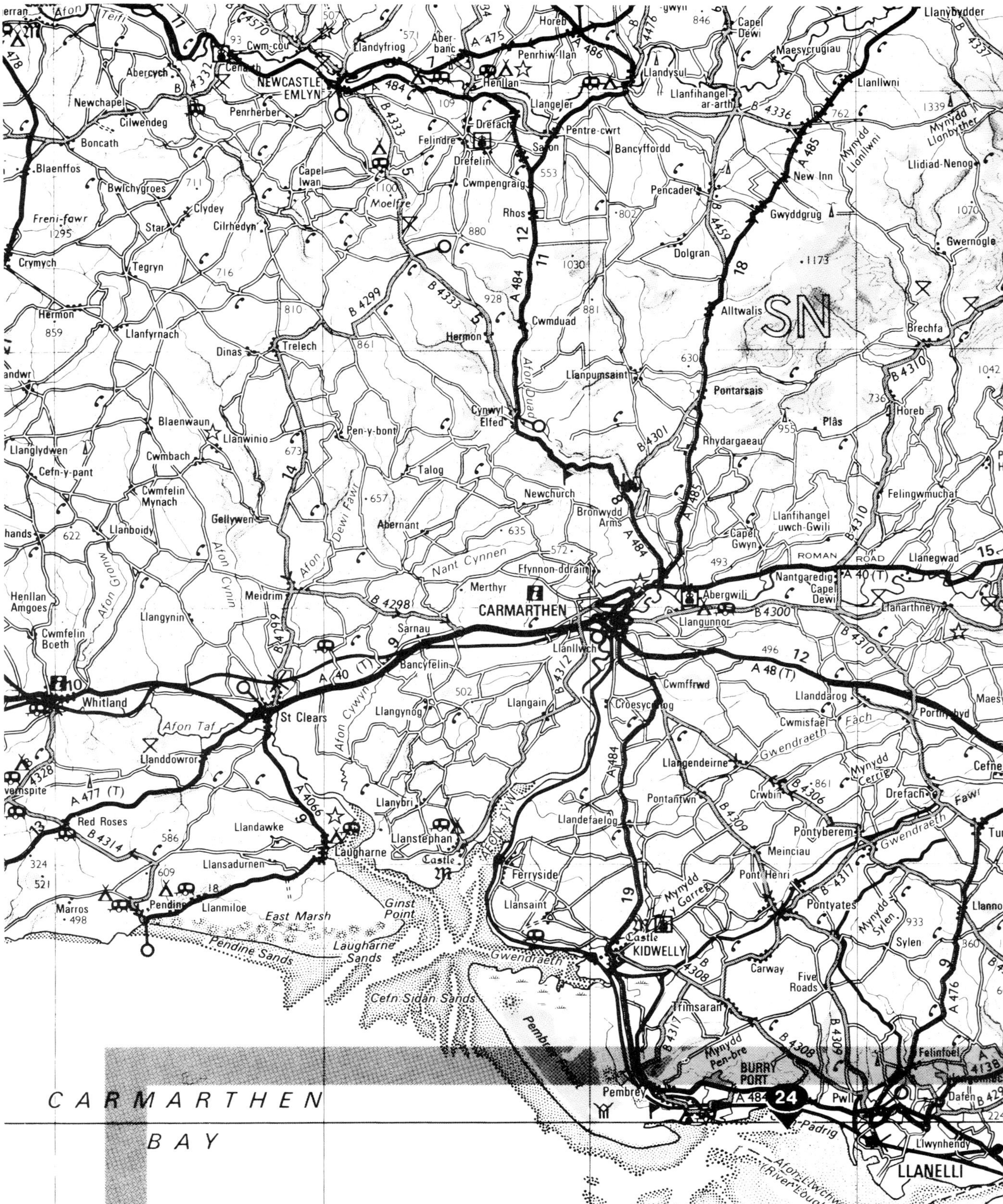

Reproduced from the Ordnance Survey 1:190 080 1989 Motoring Atlas with the permission of the Controller of Her Majesty's Stationery Office. Crown copyright reserved.

4.6

You should notice from the map that certain sections of roads are more likely to ice over than others. This is because they are the higher or more exposed parts of the route.

Also notice the location of your ICELERT sensor – at Red Roses, to the west of Carmarthen. This is quite a high spot on the main Carmarthen to Pembroke road. Certain parts of your area, however, will have slightly different conditions to those monitored by the sensor, and you should make a note of these. For example, areas further inland and higher up might freeze more easily than Red Roses.

The data from the sensors will be issued to you at 17.00 each day, and a decision has to be made quickly whether to send out the salting lorries before the crews go home at 17.30. It is not possible to call them out after this time in the activity – though, of course, they are often called out later in real life!

You should then decide, using the ICELERT data, whether any roads in your area need to be salted, and if so, which ones.

Obviously it is important to keep the major trunk roads through the area free from ice. The A40, A48 and M4 are especially important because these carry a high level of traffic, and form part of the route taken by freight traffic on its way to the ferry ports of Fishguard and Pembroke, heading for Ireland.

The other roads carry mainly local traffic, especially in poor weather conditions. People have to get to work and school, milk has to be collected from farms, and roads kept open for emergency services such as ambulances, if needed.

When you have made your decisions you should record them on a copy of the summary sheet.

You will then be told which roads did actually freeze each night and what action should have been taken. You will also be told of any 'chance events' that took place during the night – such as an accident resulting from unsalted icey roads, or an ambulance not reaching a patient quickly enough.

You can score yourself as follows:

> If you correctly decided whether to salt each road or not you should give yourself +1 mark for each route.
> If you made the wrong decisions and you salted a road that didn't freeze over, or you didn't salt a route that you should have done, then you should give yourself −1 mark for each route involved.
> If a chance event took place on an icy road you hadn't salted, then again, you should give yourself −1 mark.

The group with the highest score at the end of the activity were the best highways managers and the most skilful at interpreting the weather data!

4.6

Using the sensor data

For each night that you have to plan the salting of roads, you will be given a print-out from the ICELERT sensor at Red Roses.

The following diagram should help you to interpret the information it gives.

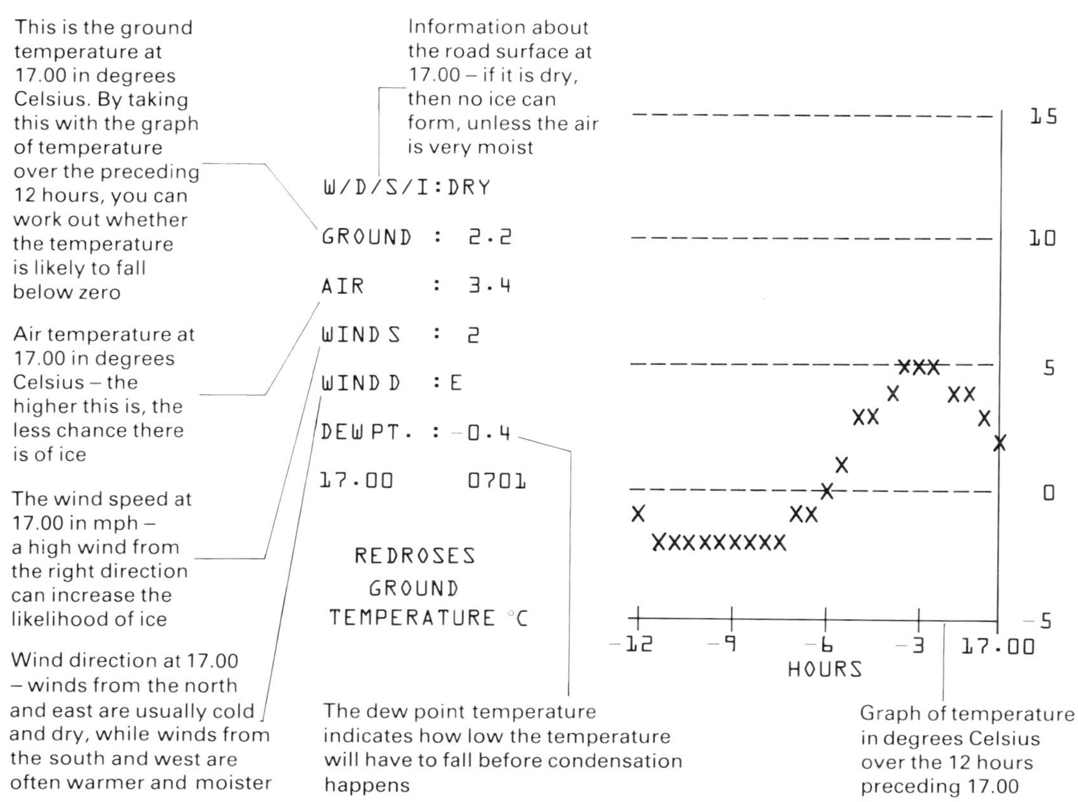

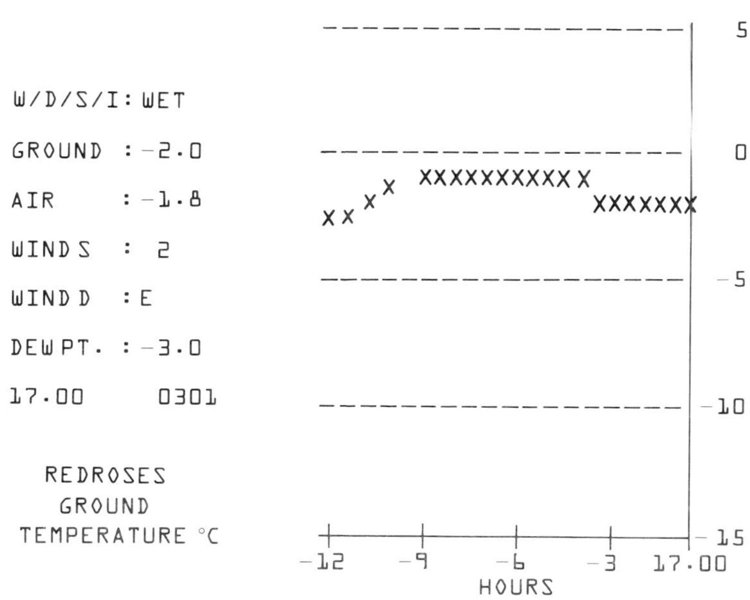

4.6

W/D/S/I: DRY
GROUND : −1.4
AIR : −0.4
WIND S : 6
WIND D : NE
DEW PT. : −2.4
17.00 0401

REDROSES
GROUND
TEMPERATURE °C

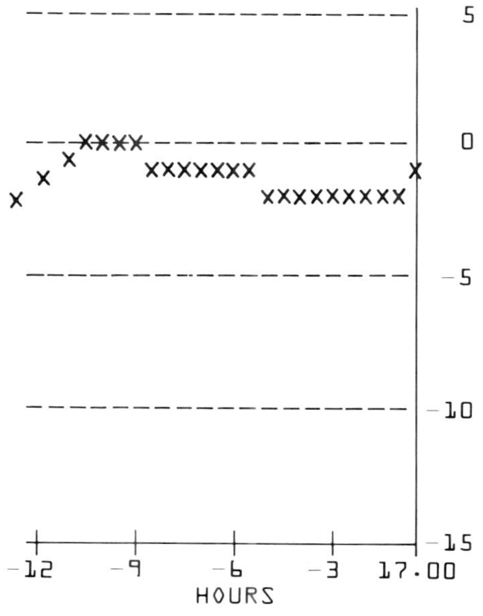

W/D/S/I: DRY
GROUND : 3.4
AIR : 4.2
WIND S : 16
WIND D : N
DEW PT. : 0.2
17.00 0501

REDROSES
GROUND
TEMPERATURE °C

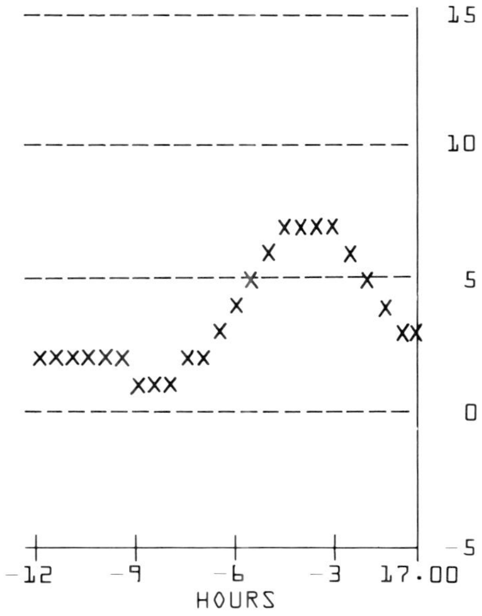

W/D/S/I: DRY
GROUND : 4.4
AIR : 3.6
WIND S : 17
WIND D : NW
DEW PT. : 0.8
17.00 0601

REDROSES
GROUND
TEMPERATURE °C

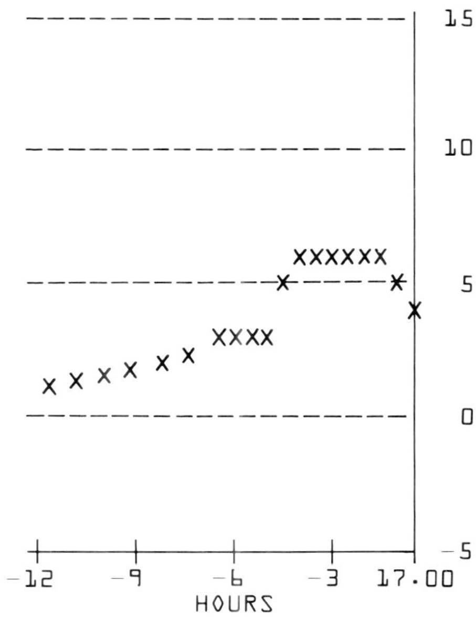

4.6

W/D/S/I: DRY
GROUND : −1.4
AIR : −1.0
WIND S : 7
WIND D : E
DEWPT. : −2.8
17.00 0701

REDROSES
GROUND
TEMPERATURE °C

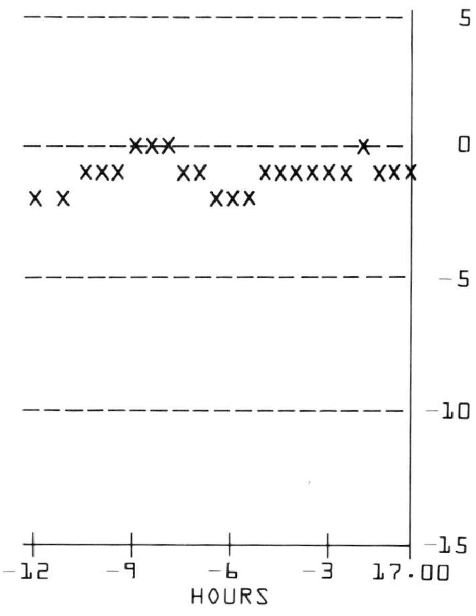

Summary sheet

Date of report	Road predictions		Actual conditions		Score
	Roads salted	Roads not salted	Roads needing salt	Roads not needing salt	

4.6

For the game organiser's eyes only

Your imagination is called for in this activity to create an exciting atmosphere. In particular, you should try to invent some events to take place each night to add to the interest of the activity. Such events might include the need for an ambulance to rescue someone hurt in a car crash on an unsalted road, the cancellation of school buses because of unsalted roads, and so on.

Here is actually what happened during each night of the activity:

3 January	Roads iced all over Dyfed.
4 January	Roads iced on higher ground only – roads were dry.
5 January	Roads did not ice anywhere.
6 January	Roads iced on higher ground only – roads were dry.
7 January	Roads iced all over Dyfed.

OBJECTIVES

CHECKLIST

4.1 High and Low Ground

Tick here

Do you

- know how temperature decreases with height? ☐

Can you

- find cold and warm areas on a satellite image? ☐
- use an atlas and a satellite image to name and identify upland areas? ☐
- draw a scatter graph to show the relationship between temperature and height? ☐

4.2 North and South

Do you

- know why temperature varies with latitude? ☐

Can you

- recognise temperature variations with latitude from a satellite image? ☐
- use an atlas and/or newspaper data to link with a satellite image pattern? ☐
- make a shaded temperature map? ☐

4.3 Distance from the Sea

Do you

- understand why temperature changes with distance from the sea? ☐

Can you

- recognise the effect of distance from the sea on temperature from a satellite image? ☐
- use newspaper weather data to link with a satellite image? ☐
- draw a bar chart and a line graph? ☐
- make a shaded temperature map? ☐

4.4 Urban Heat Islands

Do you

- understand why urban areas are warmer than surrounding rural areas? ☐
- understand how to read isotherm maps? ☐

Can you

- identify urban heat islands on a satellite image? ☐
- link atlas data and a satellite image? ☐
- carry out a group fieldwork investigation of a local urban heat island? ☐
- draw an isotherm map? ☐

4.5 Day and Night, Summer and Winter

Do you

- understand the reasons for differences between daytime and night-time temperatures? ☐

Can you

- identify differences in daytime and night-time temperatures, and in winter and summer temperatures, on satellite images? ☐
- carry out a coursework activity to find out how clouds affect temperature? ☐

4.6 Ice on the Roads

Do you

- understand the problems caused by road icing? ☐

Can you

- interpret statistical and graphical data? ☐
- use a road map? ☐
- use the map, statistical and graphical data to predict when roads are going to ice over? ☐

5.1 How and Why we Forecast the Weather

The weather is very important to many people in all sorts of ways, because it affects their lives and work. For lots of people, it is essential to know what the weather is going to be like some time in advance of it happening. For example, farmers need to know what is going to happen to the weather to be able to plan such activities as planting and harvesting crops.

The work of telling what the weather is going to be like is called *forecasting*. It is quite a difficult job because a lot of information is needed to make an accurate weather forecast.

The diagram below shows what information a weather forecaster uses to work out what the weather is going to be like, what the weather forecaster does with the information and who uses the forecasts. Today, computer programs can analyse large amounts of weather data and produce very accurate forecasts.

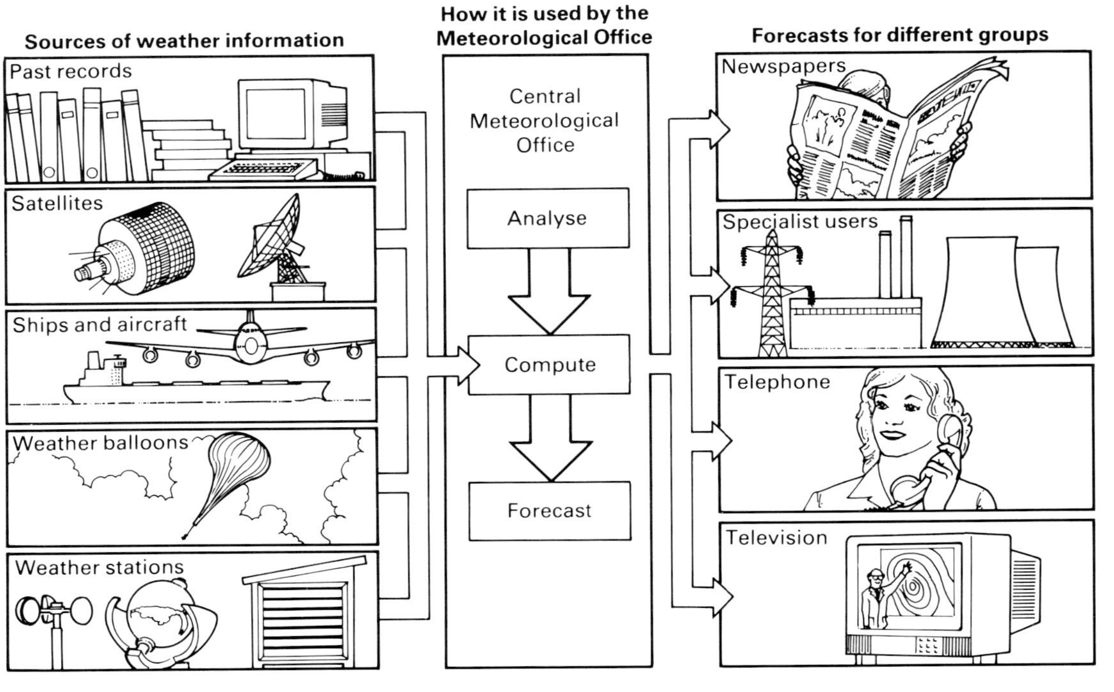

ACTIVITIES

1 Make a larger copy of the table on page 89 and fill it in to show why certain groups of people need to listen to weather forecasts. If you want to, you can add other groups to your list.

You should also try to work out how important it is for each group of people to know accurately what the weather is going to be like.

You should spend some time discussing your results with other members of the class.

5.1

Group of people	Why a forecast is important	How important is an accurate forecast?
Farmer		
Walker		
Angler		
Pilot		
School pupil		
Houseperson		

2 Look at the diagram in the first part of this section and make a list of all the main sources of weather information. Rank them in what you think is their order of importance for the weather forecaster. You should do this as a group and discuss your results with other members of the group.

3 Lots of groups of people and industries get special forecasts from the Meteorological Office. Can you think why the following need special forecasts?

British Gas
The Central Electricity Generating Board
Sealink
The Highways Department of a county council
The Royal Air Force

5.2 Who Makes Forecasts and Who Uses Them?

The work of weather forecasting in the UK is done by the Meteorological Office at Bracknell in Berkshire.

The Meteorological Office is responsible for collecting weather information from many different sources, including weather satellites, weather ships and land-based weather stations. The information collected is then processed using expensive and complex computer technology, and weather forecasts are produced for many different kinds of user.

Short-range forecasts are for a day or two ahead and are usually quite accurate. These are the forecasts you see on television and read in newspapers. You can also phone up for a weather forecast.

Longer-range forecasts try to predict what the weather is going to be like for a few days ahead. These forecasts are used not only by the general public but also by specialist users, such as farmers and industrialists like British Gas and the CEGB. The shipping forecast is perhaps the best known specialist forecast – it can be heard on Radio 4.

ACTIVITIES

You should carry out both the following activities as a class – or perhaps they might form the basis for a coursework item.

1 Try to find as many media sources of weather forecasts as you can, and make a list of where weather forecast information can be found. You could make a chart similar to the one below, which has been partly filled in for you.

Source of forecast	Area covered by forecast	Detail in forecast	Frequency of forecast
Radio 4	Sea areas around the UK	Pressure, wind speed and direction, visibility	Four times a day
Western Mail	Wales	Temperature, rainfall, wind speed and direction	Daily

5.2

2 How much notice do people take of weather forecasts?

Why not do a survey, either just in school, or with members of your family or in your local community, to find out where they get their weather information, and how much use they make of it? You could find out their attitudes to different kinds of weather as well, although you don't want to make your questionnaire too long.

You could use the questionnaire below as a model.

Weather survey

Location of survey ____________________

1 Occupation ____________________

2 How often do you read/listen to/watch a weather forecast?

More than twice each day
Daily
Every one or two days
Weekly
Only occasionally

3 Which weather forecast do you use?

Newspaper	**If so, which one?** ____________
Radio	**If so, which station?** ____________
Television	**If so, which channel?** ____________
Others	**Please say which.** ____________

4 If you watch a television forecast, which part of the forecast do you take most notice of?

Satellite image
Weather maps – picture map
– isobaric weather map
Description by the forecaster

5 How do you think that weather forecasts could be made more useful?

____

6 Do you organise your day's activities around the weather forecast information?

Yes **Please give a few examples.** ____________________

____

____

No

7 How often do you think that the weather forecasters get it right? Please give a score out of ten.

______/10

Thank you for answering our questions. If you want to know about the results of our survey, please give us your address and we will post them to you.

5.3 Making your Own Forecasts (1)

Making your own weather forecasts can be fun. If you forecast the weather correctly, they can also be useful!

There is, however, some basic information you will need before you can have a go at making a decent forecast.

1 You will need the current satellite image which shows the UK and as much of the north Atlantic Ocean as possible. Most of the weather which the UK has comes from the west and so you need to know what weather is on its way. If you have an image which only shows the UK and most of Europe, you'll usually only be able to say what weather we have just had, not what weather we are going to have!

2 You will also need to know how fast and in which direction the weather is heading towards the UK. Although most weather systems from the Atlantic will affect the UK, it is important to know as accurately as possible when and where they do this.

You have already had a lot of practice at working out what the weather is like using a satellite image, so **1** should be no problem. There are a number of ways of working out the speed and direction on a weather system is moving in. Four different ways are explained below and on pages 93–4, and it is important that you try doing these before attempting your own forecast.

ACTIVITIES

1 At certain times during the day, Radio 4 broadcasts the *shipping forecast*. This is a detailed description of the weather in each of the sea areas around the UK.

The forecast always starts off with a general description of the weather systems, saying where they are at the moment, and where they are going to be 24 hours later.

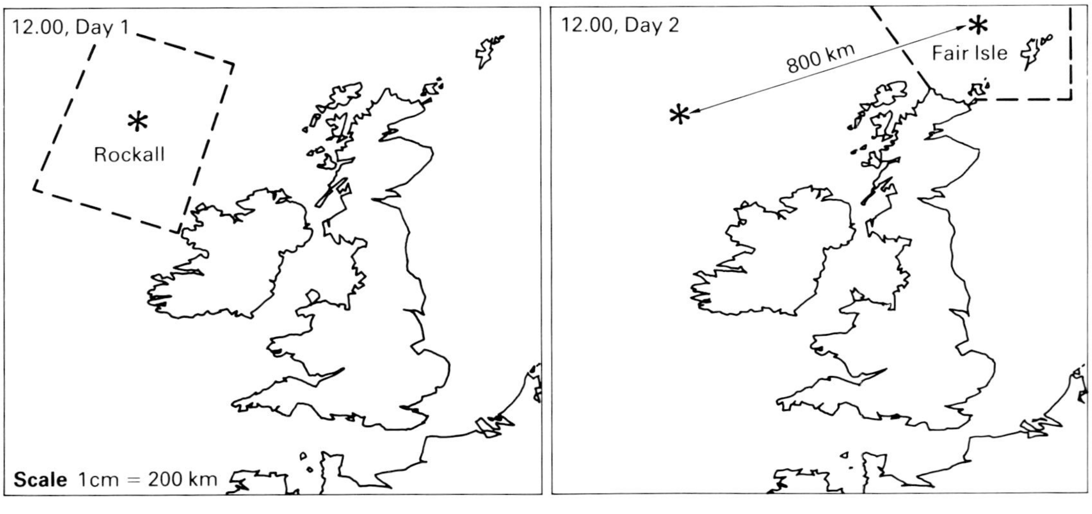

The low pressure has moved 800 km in 24 hours. The speed equals $\frac{800}{24}$ = 33 km an hour, in a NE direction.

5.3

A typical statement might be: 'The low pressure system in sea area Rockall will move slowly north-west, expected Fair Isle by midday tomorrow' (forecast at 12 noon).

If you put this information on a copy of the shipping forecast map in Appendix 4, marking the position of the low pressure at the time of the forecast and 24 hours later, you should be able to calculate the speed and direction the weather is moving in.

2 Another source of information about the movement of weather systems is the Meteorological Office's 'Weathercall' service. This provides a detailed forecast for different areas of the UK.

If you use this service, don't listen to the whole forecast, otherwise you'll be able to cheat when making your own! Just listen to the first part, when details are given about the movement of pressure systems.

Then you can repeat the last part of Activity 1 to work out the speed and direction of movement of the weather systems.

3 It is possible to work out the speed and direction at which weather is heading towards the UK using two satellite images taken at different times.

Obtain two images that have been taken more than two hours apart. (If you try using two Meteosat images taken any less than two hours apart, you may find it difficult to spot any movement of the cloud patterns.)

For both images, either get a hard copy or make an accurate map of the major cloud patterns over the Atlantic Ocean approaching the UK. If you can find a front then it is a good idea to focus your attention on this.

By comparing the position of the clouds at the two different times, it should be possible to work out how fast the clouds are moving, and in which direction.

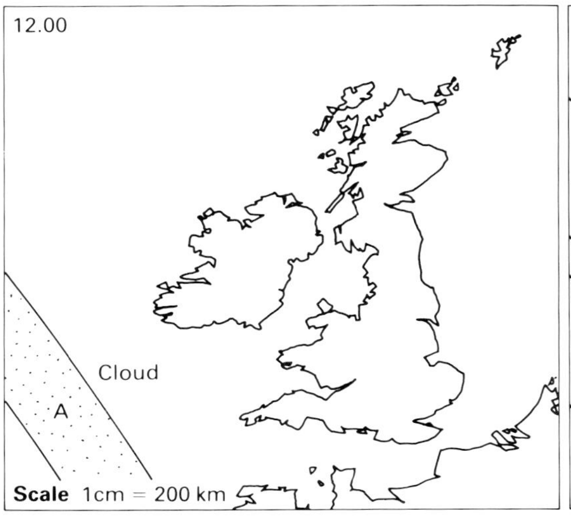

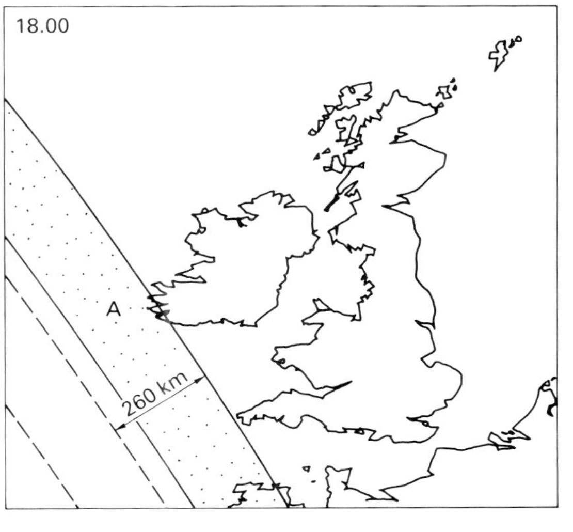

Cloud band A has moved 260 km in 6 hours. The speed equals $\frac{260}{6}$ = 43 km an hour, in a NE direction.

5.3

4 A final source of information about the speed and direction of the movement of weather can come from weather maps published in newspapers.

Some weather maps show the whole of the north Atlantic. You should choose two maps, one day after the other, and pick out a depression or front that is moving towards the UK.

You should measure the distance the fronts have travelled over the 24 hours that separate the two maps, and calculate the speed and distance of movement.

It is also possible to work out how fast the centre of low pressure is moving, rather than the fronts.

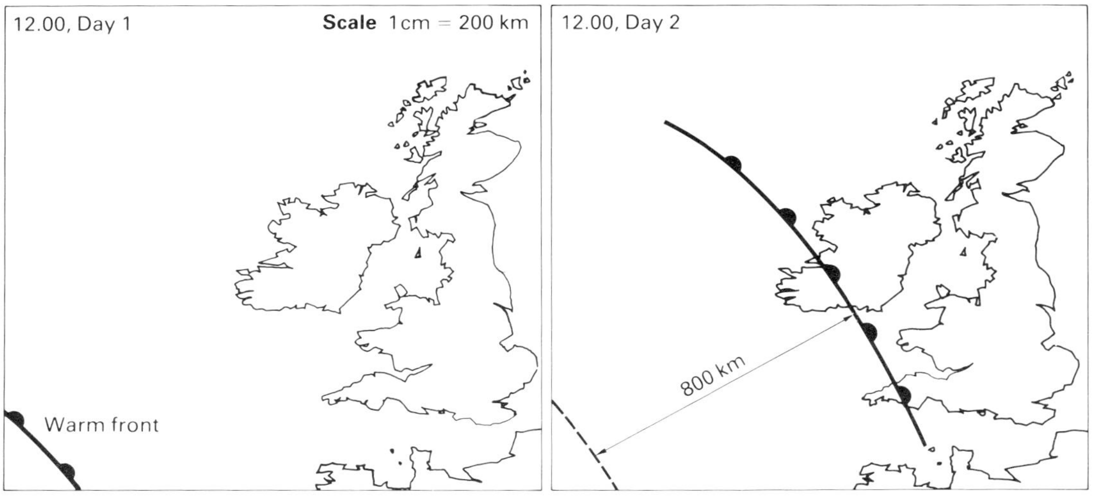

The warm front has moved 800 km in 24 hours. The speed equals $\frac{800}{24}$ *= 33 km an hour, in a NE direction.*

5.4 Making your Own Forecasts (2)

Making a forecast isn't difficult – the problem is in getting it right!

To forecast the kind of weather the UK is going to have you need to know:

1 what the weather is like at the moment, both over the UK and immediately to the west over Ireland and the Atlantic Ocean.

2 how fast and in what direction this weather is moving towards, or across, the UK.

You can then work out where the weather is going to go over the next 24 hours or so – always bearing in mind that weather patterns change very quickly. Fronts, for example, may speed up or slow down suddenly, affecting the accuracy of your forecast.

COURSEWORK ACTIVITIES

1 To forecast the weather you should follow these steps.

a Look at the current satellite image and draw a map to show the pattern of cloud. It is helpful to begin with if you concentrate on the clouds that are likely to bring rain.

b Then work out how quickly these clouds are moving towards and across the UK.

c The next move is to mark on your map where the clouds are likely to be in, say, 24 hours' time, if they carry on moving at the same speed over the UK.

d You can now write a forecast of what the weather is going to be like over the next 24 hours. Your forecast should include:

a general description of the weather now;

a simple description of what is going to happen over the next 24 hours; and

if possible, a region by region weather forecast.

2 After you have made your forecast, you could make either an audio or a video tape, and then perhaps compare it with the radio or television forecast.

3 Weather map symbols have been provided with this resource and activity pack. Perhaps they could be used on a large map of the UK to give a daily weather forecast, to be displayed outside the geography room.

4 You could also try to find out how many of the common weather sayings are true – such as 'Red sky at night, shepherd's delight; red sky in the morning, shepherd's warning'.

5.5 Did you Get your Forecasts Right?

Many people say that the weather forecasts are not very accurate and that they could do better themselves! This isn't really true – nine times out of ten the forecasts produced by the Meteorological Office are basically correct.

However, if you want to, you can try to beat that record by carrying out the activity below.

ACTIVITY

This activity can either be carried out as a class group, or perhaps as a coursework activity to do on your own.

a Choose a number of different sources of weather forecasts – including, of course, your own!

b On a large copy of the diary on page 97, fill in what each forecast said would happen to the weather in your area each day for a week or so.

c At the end of each day, make a note of what *actually* happened to the weather.

d Then give each weather forecast a score out of ten. To do this, you should count the number of pieces of weather information given in the forecast and then make a note of the number of accurate predictions. For example, if the forecast gave five pieces of information about the weather, and three turned out to be correct, then this would be a score of $\frac{6}{10}$.

One point to remember, though, is that the more information given by the forecast, the more likely it is to be slightly wrong.

e Rank the forecasts in order of accuracy. Which was the most accurate for your local area?

5.5

Weather forecast recording sheet

You will need one sheet for each day you compare the weather forecasts.

Location ______________________________

Date ______________________________

Source of forecasts	The forecast weather	Score and rank
School forecast		
Radio		
Newspaper		
Television		
Phone		
Actual weather		

5.6

Farming and the Weather

The weather has a great effect on the day-to-day activities of farmers. The activity that follows involves you putting yourself in the position of a farmer at harvest time.

ACTIVITY

You have five fields of spring barley to harvest. Your barley was planted in the spring and can be harvested any time during 14–28 August. You cannot harvest before this as it will not be ripe enough, and if you leave your harvest any later the over-ripe grain will begin to fall to the ground and you will lose a lot of your crop.

Harvesting a crop may sound easy, but barley can really only be harvested in dry weather. If the grain is at all wet then the farmer either has to risk it going rotten in storage, or has to spend money on drying it. A wet field also poses problems for heavy combine harvesters, which could run the risk of getting stuck. Even damp weather can be a problem.

This is how you should carry out the activity.

a You should get into groups of three or four and each group should be given a summary sheet and a set of weather cards. One member of the group should record the decisions you have made on the summary sheet, and the weather cards should be placed face down in front of you.

b Take the first weather card. Each card has a weather map for 6 a.m. on it, together with a short Meteorological Office summary of what the weather is going to do. Look carefully at all the information on the weather map, and make a basic weather forecast. You need do no more than work out whether it is going to be rainy or dry. Then, on the basis of your forecast, decide whether you are going to harvest any of your fields. You can only harvest one field a day, and after a period of rain, you have to let the field dry for a day before you can harvest it.

c At the end of each 'day', the person in charge of the activity will tell you what actually happened to the weather. Bonus points will be given if you have predicted the weather well and have managed to harvest a field, and penalty points will be given if you predicted the wrong weather and harvested on a wet day. The group with the best scores at the end of the month will be the best farmers!

d There is a 'chance' card as well, which might ruin your harvest. This involves heavy rain which can easily lay a field of barley flat, and make it impossible to harvest. This card will be read out at some point during the activity.

5.6

Summary sheet

Tick the field numbers when you have harvested them: 1 2 3 4 5

Date in August	Predicted weather	Planned activity	Actual weather	Score
14				
15				
16				
17				
18				
19				
20				
21				
22				
23				
24				
25				
26				
27				
28				

Total score ____________

Weather maps

The maps are for the situation at 6 a.m. each day from 14 to 28 August 1979. They should be photocopied on to card, and cut up to make playing cards – one map per card. Each group should be given a set of cards in date order.

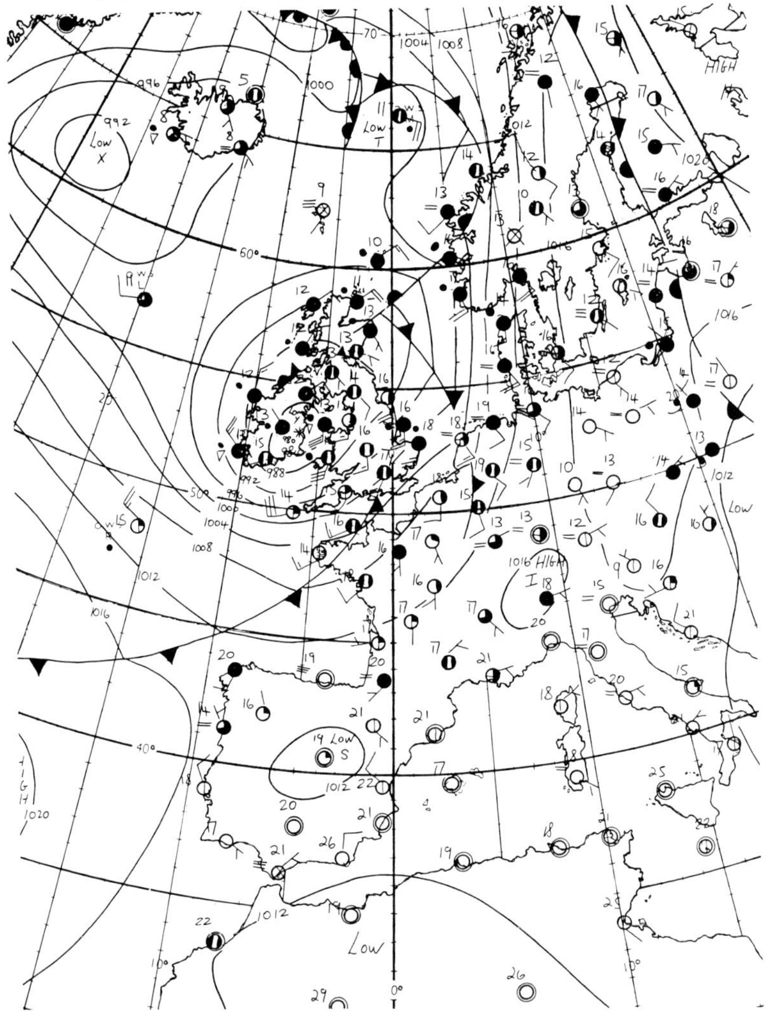

14 August – during the next 24 hours, the depression will continue to move north-north-east over Ireland and Scotland, with associated frontal troughs moving north and east over much of Britain, followed by a showery north-westerly airstream. (Crown copyright reserved)

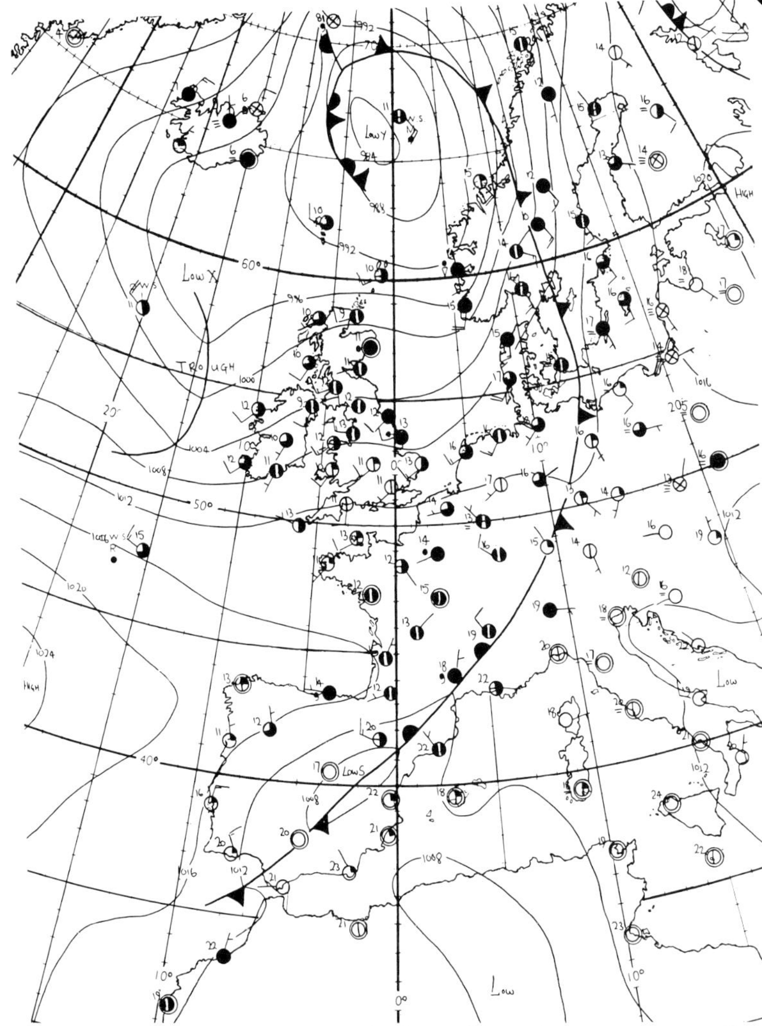

15 August – during the next 24 hours, the low will continue to move north and fill. A deepening depression in mid-Atlantic is expected to move fairly quickly south-east to be off south-west Ireland by tomorrow morning, with renewed gales likely in the Southwest Approaches. (Crown copyright reserved)

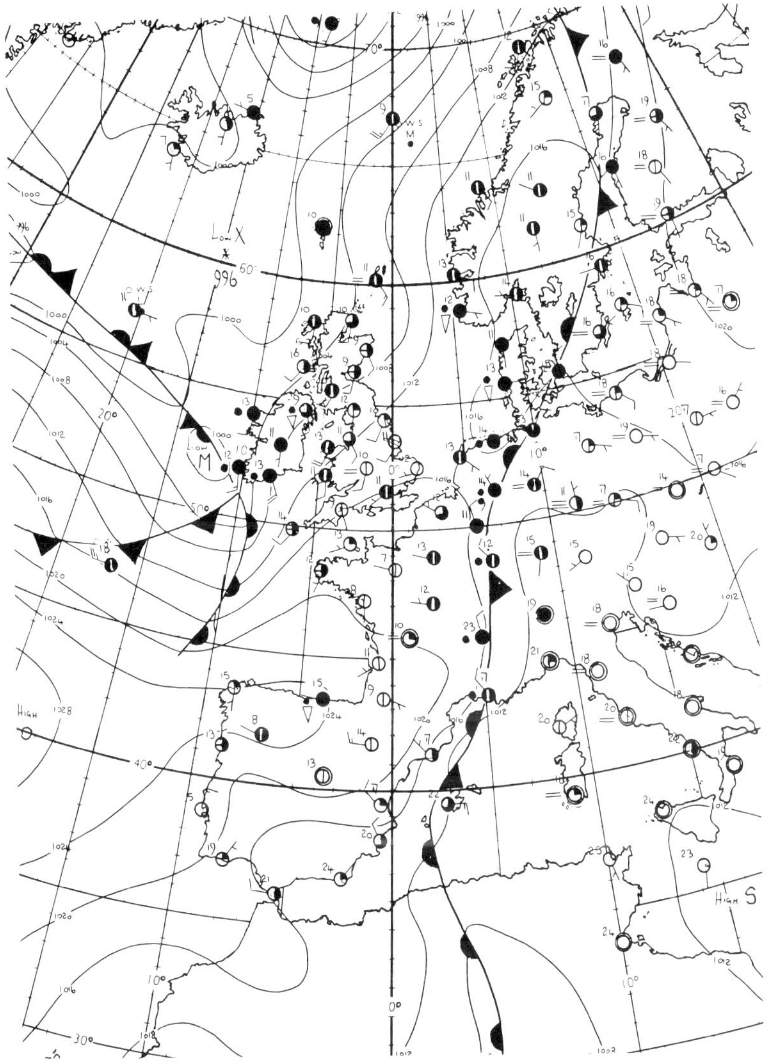

16 August – during the next 24 hours, a deepening depression associated with this system will move east into the English Channel, with the fronts affecting most of southern Britain.
(Crown copyright reserved)

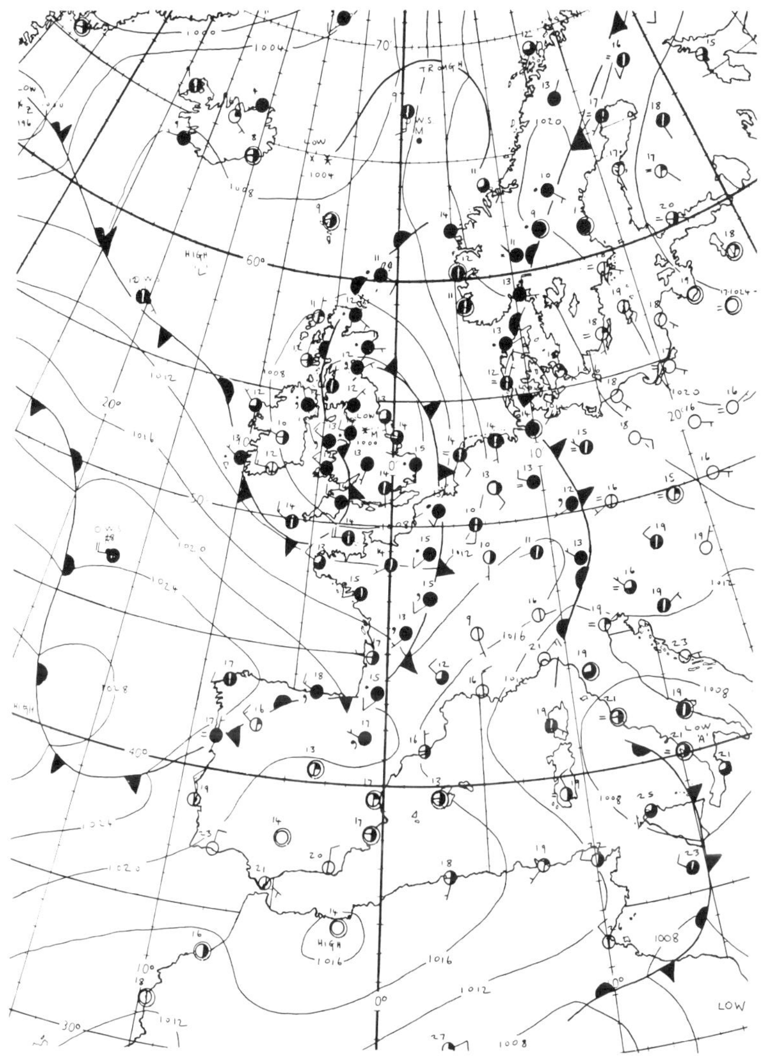

17 August – during the next 24 hours, the depression is expected to remain slow-moving off the coast of north-east England, with frontal troughs continuing to affect northern Scotland. A shallow wave depression is expected to move south-eastwards affecting parts of south-west England for a time.
(Crown copyright reserved)

5.6

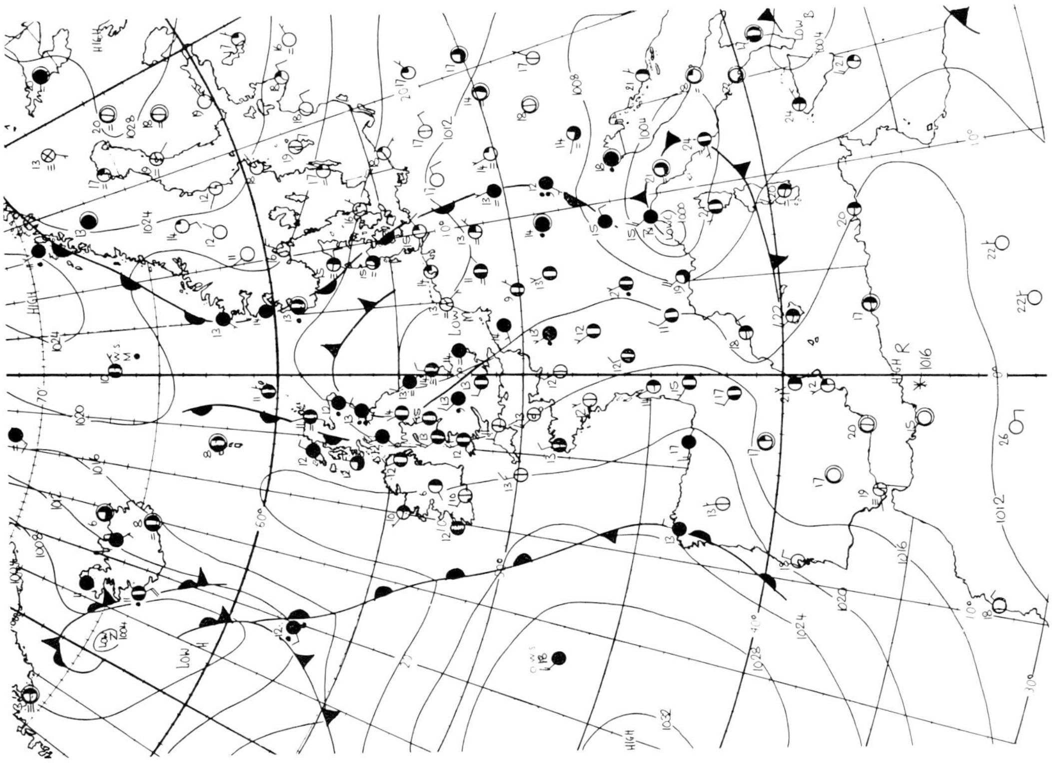

18 August – during the next 24 hours, the depression is expected to fill completely whilst a large circulation develops around a low over the western Mediterranean. The area of high pressure will establish itself over Scotland, as an occluded front approaches western coasts. *(Crown copyright reserved)*

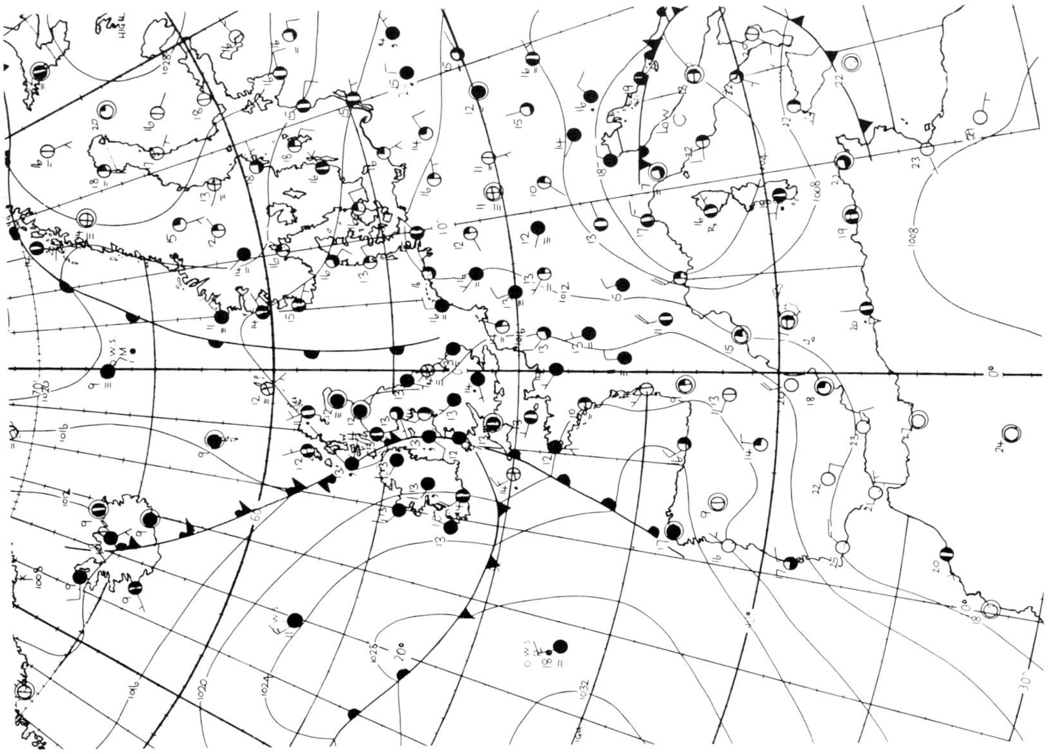

19 August – during the next 24 hours, the depression over the Irish Sea will move south-east and fill, and the associated frontal trough is expected to clear the east coast tonight. This will be followed by the ridge of high pressure, but a developing depression over the central Atlantic moving east is expected off western Ireland by tomorrow morning. *(Crown copyright reserved)*

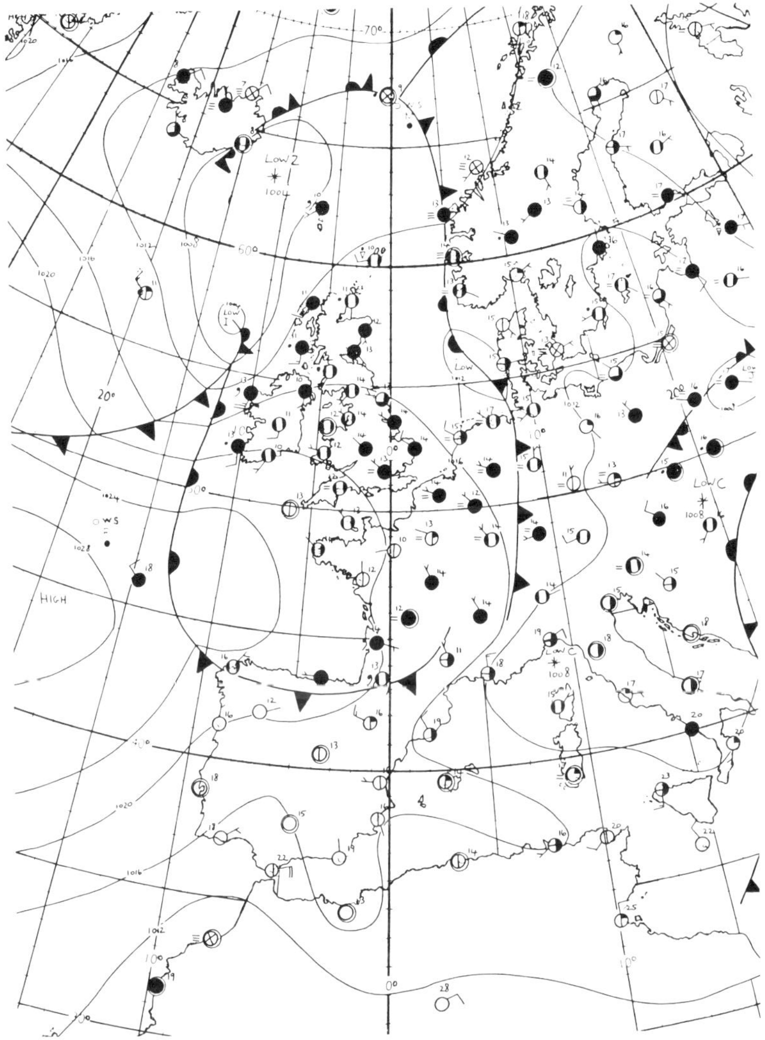

20 August – during the next 24 hours, this depression will deepen further and is expected to reach the Shetland area, with its associated frontal troughs moving east to cross most parts of the British Isles.
(Crown copyright reserved)

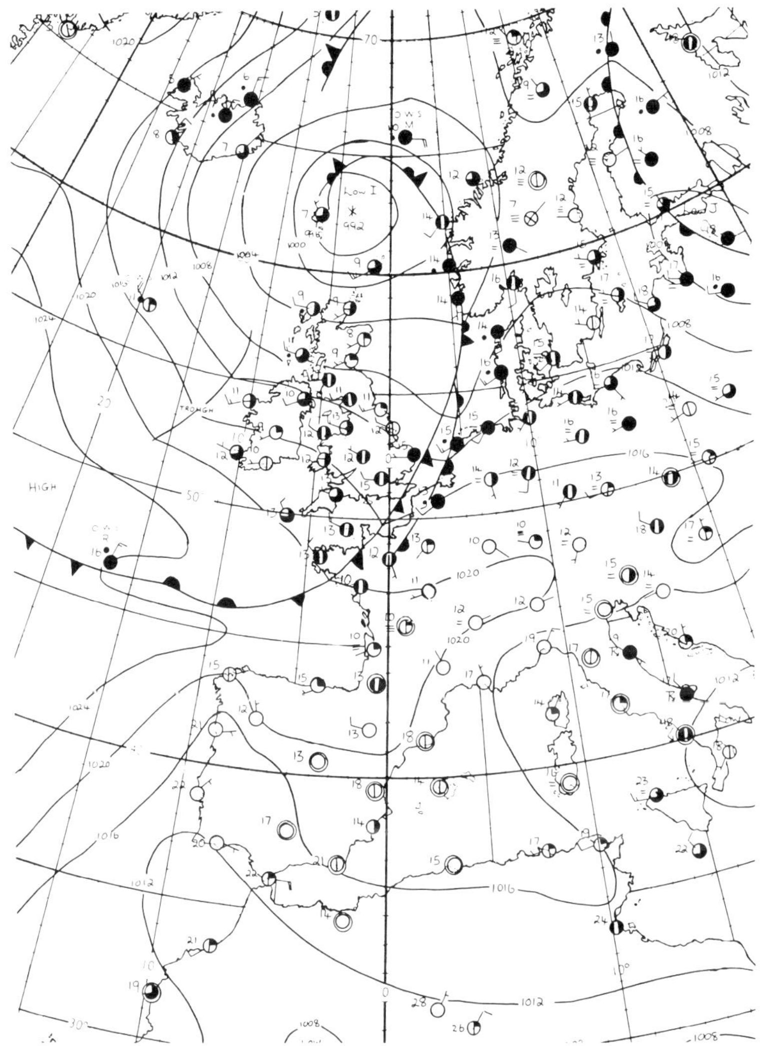

21 August – during the next 24 hours, the low is expected to remain slow-moving to the north of Scotland, maintaining a west to north-westerly airstream over the British Isles.
(Crown copyright reserved)

5.6

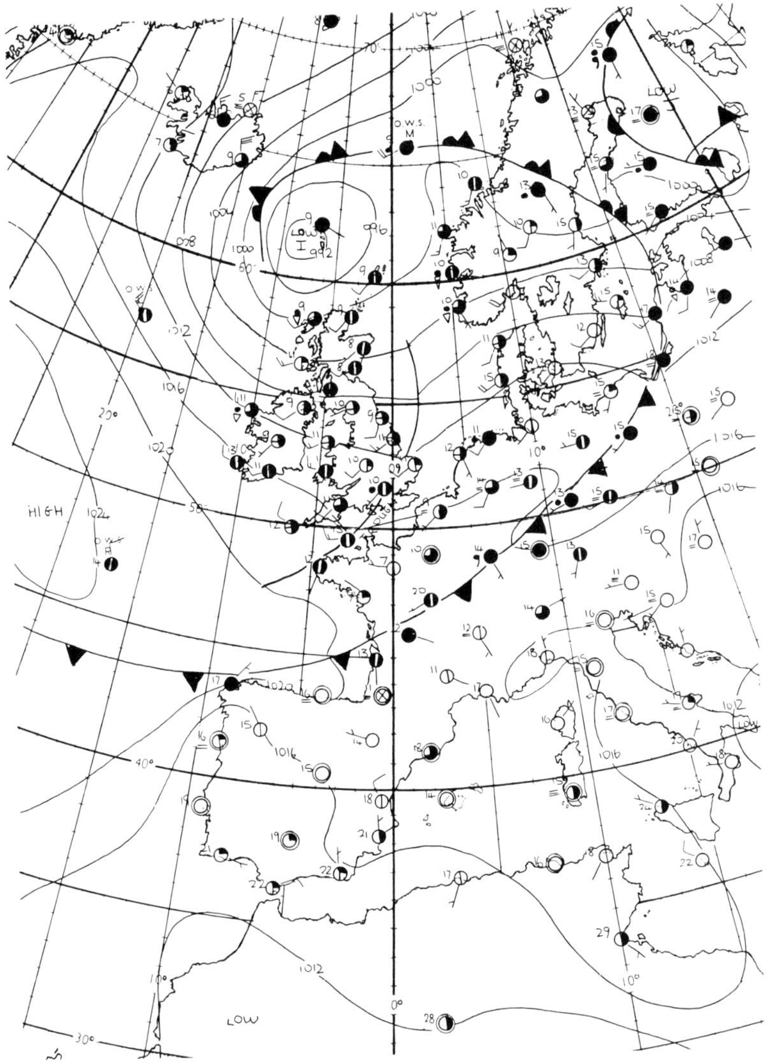

22 August – during the next 24 hours, the depression is expected to move slowly south-east towards Shetland, maintaining the unstable north-westerlies over the British Isles.
(Crown copyright reserved)

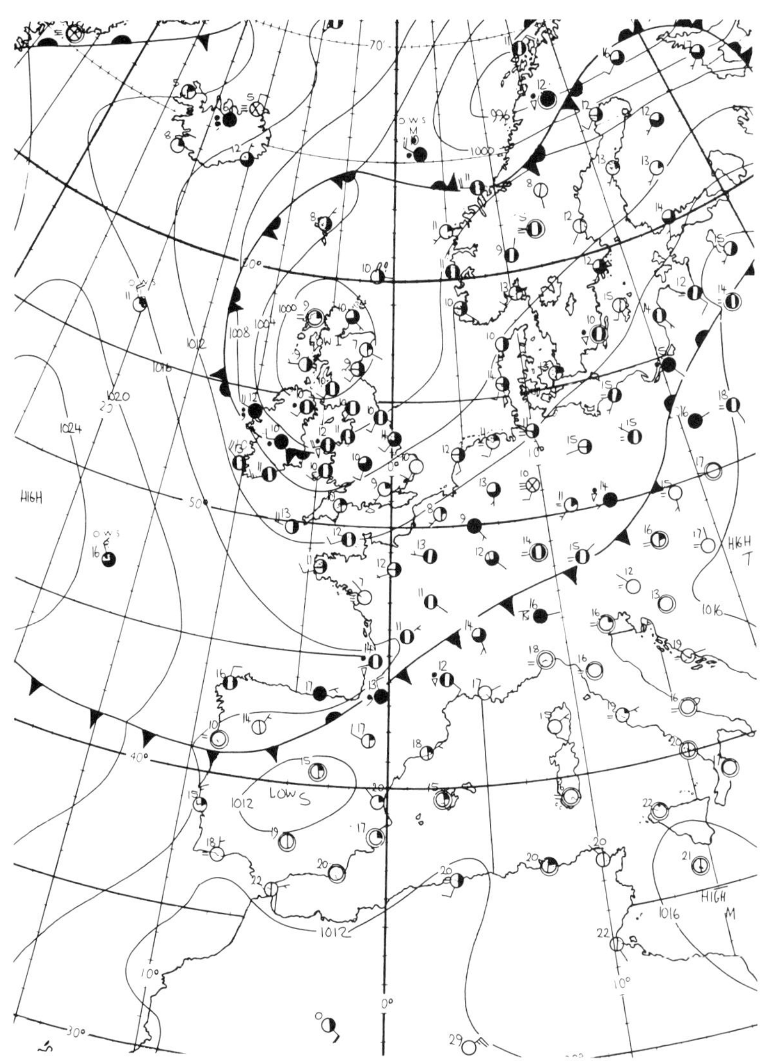

23 August – during the next 24 hours, the depression will continue to move south-east across Scotland and north-east England, with the north-westerly airflow persisting over the British Isles.
(Crown copyright reserved)

5.6

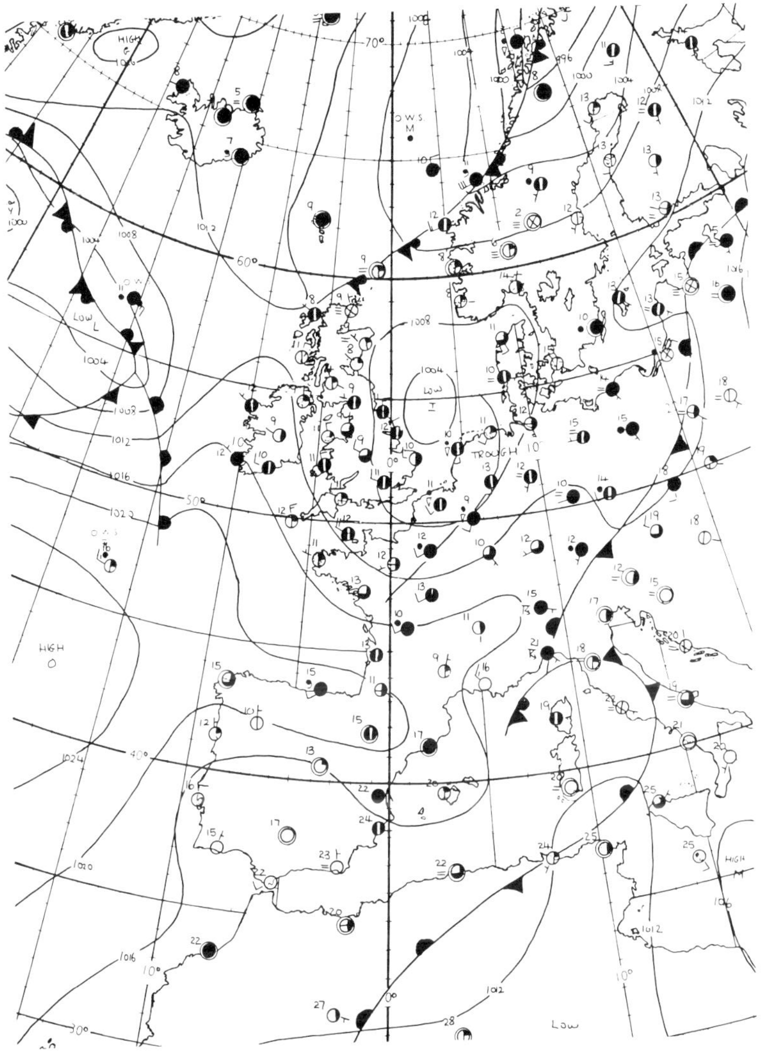

24 August – during the next 24 hours, this depression will move east into the Continent whilst a further depression moves south-east from the Atlantic into the south-west of England.
(Crown copyright reserved)

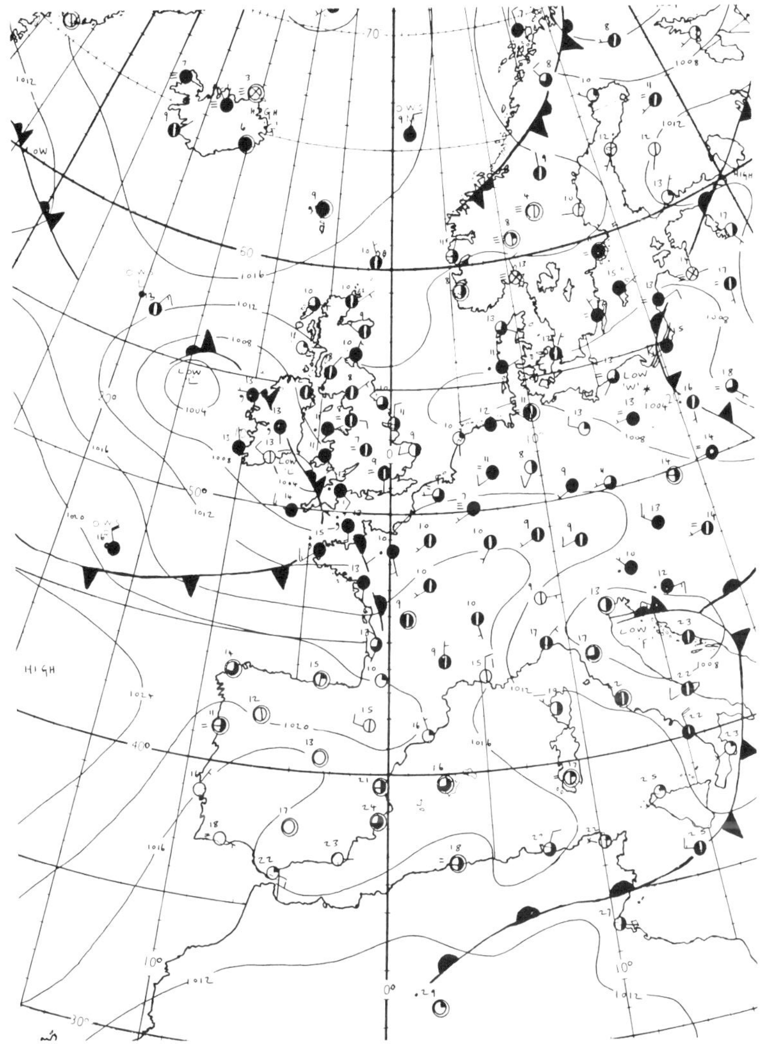

25 August – during the next 24 hours, the low over south-west England is expected to move away east-south-eastwards, as a weak ridge of high pressure builds over Scotland.
(Crown copyright reserved)

5.6

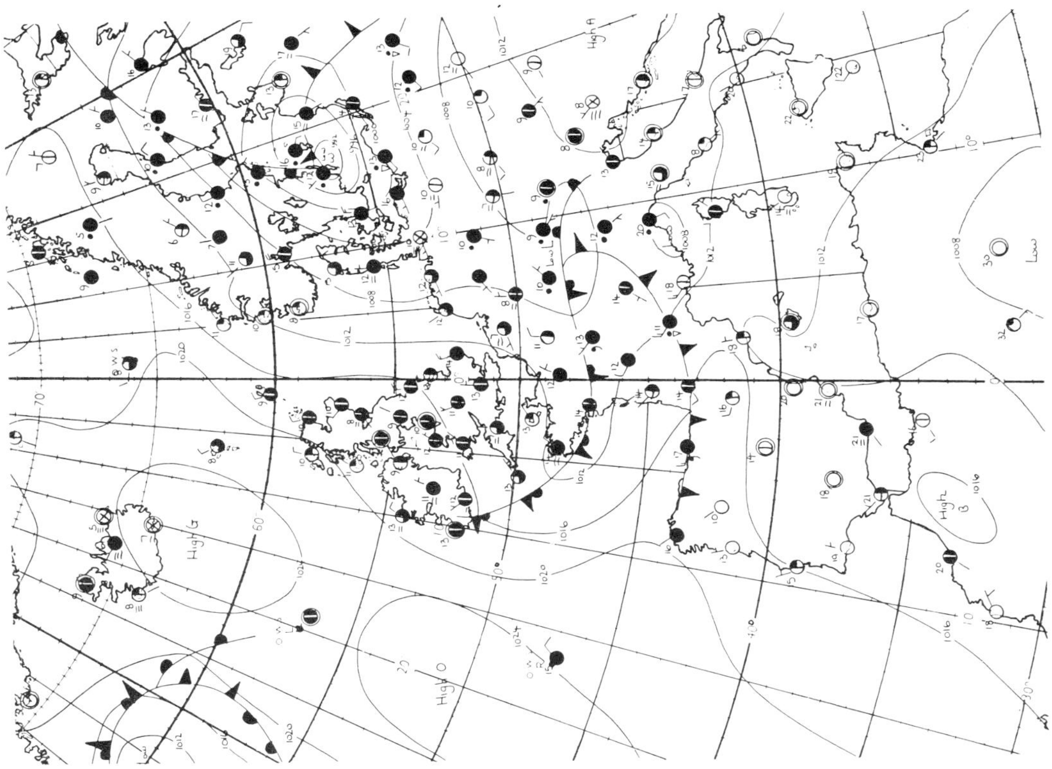

26 August – during the next 24 hours, the depression over France is expected to continue moving eastwards, but a northerly flow will be maintained over the British Isles, as a depression over south Sweden moves slowly north-west and a ridge of high pressure persists over and to the west of Ireland. (Crown copyright reserved)

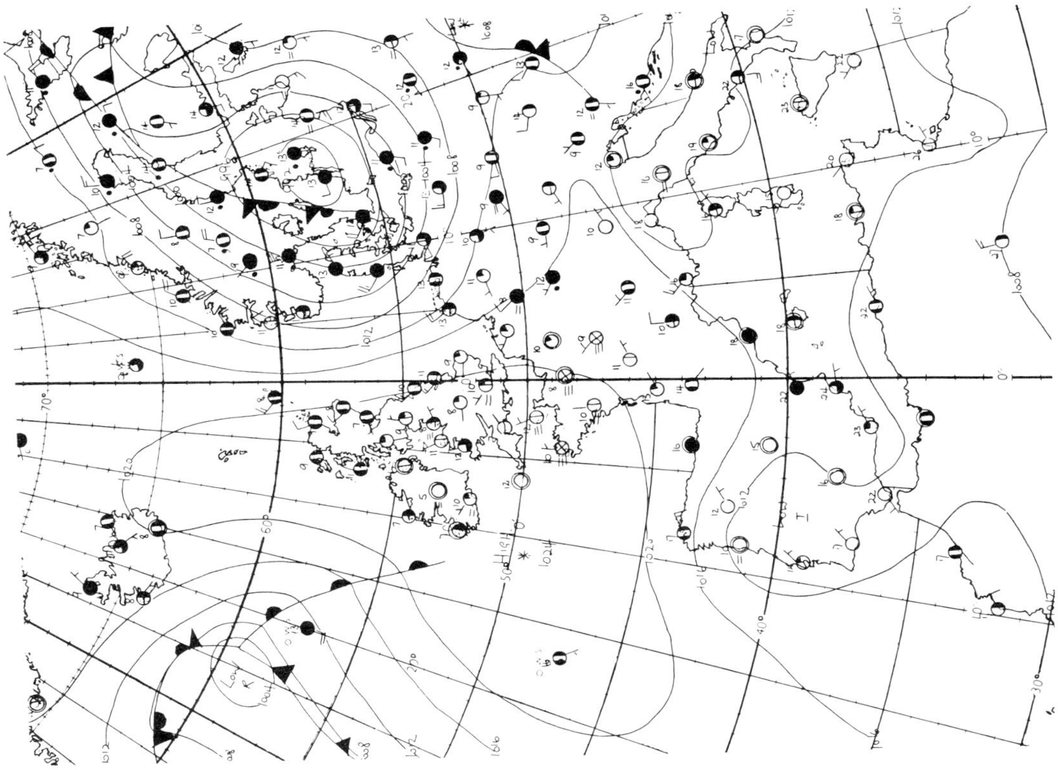

27 August – during the next 24 hours, the anticyclone is expected to reach Brittany although the Swedish depression will move only slowly east. Over northern Britain the ridge will be eroded as an Atlantic low approaches southern Iceland. (Crown copyright reserved)

5.6

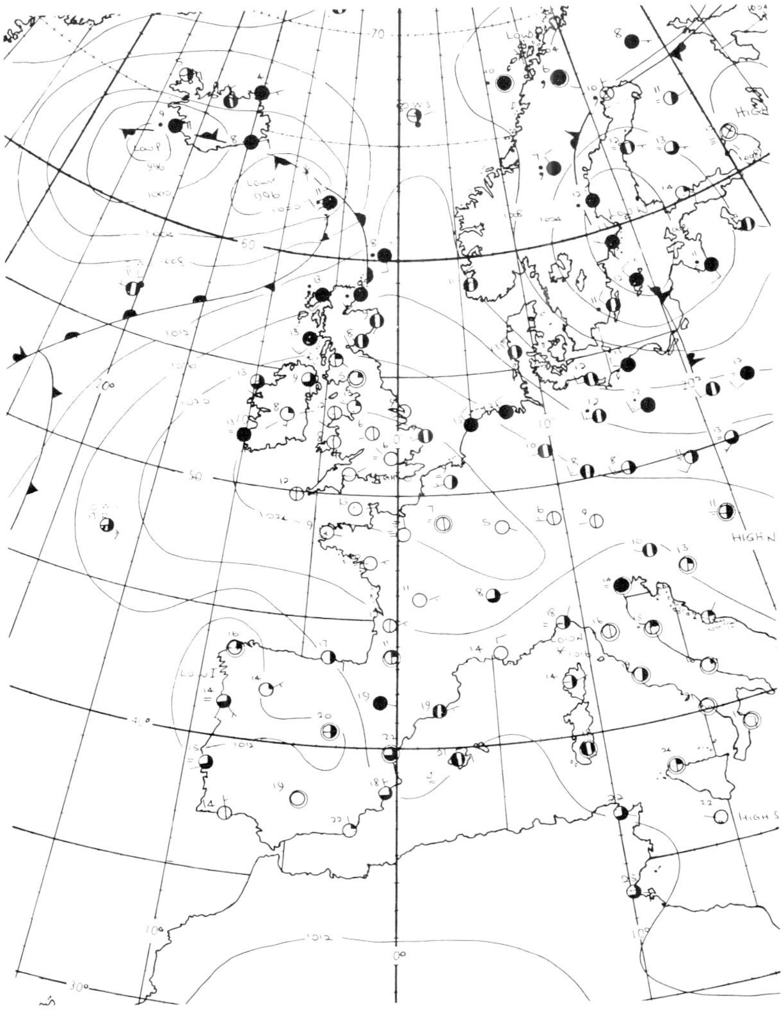

28 August – during the next 24 hours, the anticyclone is expected to move slowly east-south-east whilst fronts affect north-western districts.
(Crown copyright reserved)

5.6

Actual rainfall data

For the game organiser's eyes only!

Date	Rainfall (mm)	Points if harvest is made that day	Comments
14	1.0	−10	Wet – grain would need a lot of drying
15	0.0		No harvest possible – first day after rain
16	10.0	−10	Wet – grain would need drying
17	0.0		No harvest possible – first day after rain
18	0.0	+5	Harvest possible
19	3.0	−5	Wet – grain would need drying
20	2.0	−5	Wet – grain would need drying
21	0.0		No harvest possible – first day after rain
22	1.0	−5	Wet – grain would need drying
23	3.0	−5	Wet – grain would need drying
24	5.0	−10	Wet – grain would need a lot of drying
25	0.0		No harvest possible – first day after rain
26	0.0	+5	Harvest possible
27	0.0	+5	Harvest possible
28	0.0	+10	Harvest possible – grain really dry

Chance card

Read this out on the specified day:
24 August: rain flattens part of Field 4 – if groups have not harvested this field they 'earn' −2 marks.

OBJECTIVES CHECKLIST

5.1 How and Why we Forecast the Weather

Do you — Tick here

- understand why forecasting the weather is important? ☐
- understand that the weather affects different groups in different ways? ☐

5.2 Who Makes Forecasts and Who Uses Them?

Do you

- understand that there are different kinds of weather forecast? ☐

Can you

- analyse media sources of weather forecast information? ☐
- carry out a questionnaire, and analyse the results to find out how often people listen to and take notice of weather forecasts? ☐

5.3 Making your Own Forecasts (1)

Can you

- calculate the speed of movement and direction of movement of a weather system from:
  - the shipping forecast? ☐
  - a telephone weather forecast? ☐
  - two satellite images? ☐
  - two media weather maps? ☐

5.4 Making your Own Forecasts (2)

Can you

- use satellite images to make a weather forecast? ☐
- record this forecast on either audio or video tape? ☐
- investigate the truth of folklore sayings? ☐

5.5 Did you Get your Forecasts Right?

Can you

- collect weather forecast information from a number of different sources? ☐
- compare different forecasts in terms of their accuracy to predict the weather? ☐

5.6

Farming and the Weather

Do you

- understand the issues involved in deciding when to harvest a barley crop? ☐

Can you

- interpret Meteorological Office weather maps for a period of time? ☐
- use Meteorological Office weather maps to predict suitable conditions for harvesting a crop? ☐

Section / Skill	Using visible images	Using infra-red images	Linking images	Observation	Description	Research	Calculation	Classification	Matching	Ranking	Prediction	Empathy	Using: Graphs	Using: Diagrams	Using: Media weather data	Using: Text	Using: Atlas/maps	Making: Graphs	Making: Diagrams	Making: Text	Making: Maps	Making: Sketches	Practical experiments	Weather measurement	Model making	Visual presentation	Coursework activity
1.1						√																					
2.1	√												√	√							√						
2.2		√											√	√				√			√						
2.3	√	√	√											√							√					√	
2.4	√	√	√											√		√					√		√				
2.5	√	√		√	√													√	√		√					√	√
2.6	√	√				√															√	√					
2.7	√	√													√		√				√						
2.8	√	√												√							√						
2.9						√					√	√			√	√											
3.1	√	√					√														√						
3.2	√														√		√				√						
3.3	√	√												√	√						√						
3.4	√	√						√													√			√			
3.5	√	√		√	√												√	√	√		√		√	√	√	√	√
3.6	√																				√			√	√		
3.7	√	√		√	√												√	√			√			√		√	√
4.1		√															√				√						
4.2		√												√	√		√				√						
4.3		√													√		√				√						
4.4		√													√		√	√	√		√			√		√	√
4.5		√		√	√													√			√						√
4.6		√		√	√				√		√	√									√						
5.1	√	√								√		√		√													
5.2	√	√								√		√						√									√
5.3	√	√					√														√						
5.4	√	√					√				√									√	√					√	
5.5	√	√					√				√							√			√			√			√
5.6											√	√															

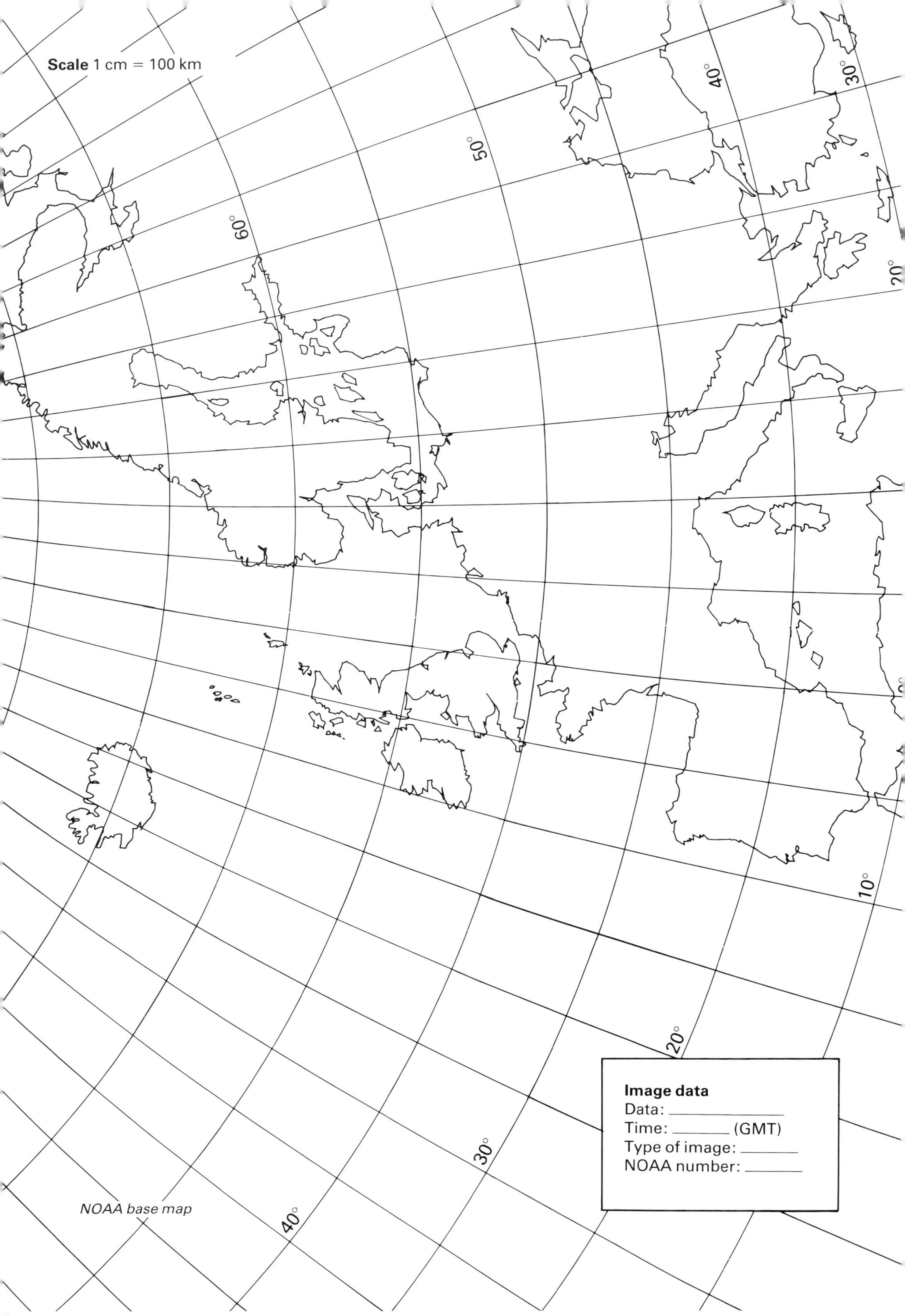
Scale 1 cm = 100 km
60°
50°
40°
30°
20°
10°
20°
30°
40°
Image data
Data: ________
Time: ______ (GMT)
Type of image: ______
NOAA number: ______
NOAA base map

Image data **Scale** 1 cm = 100 km

Date: ______________ Time: _______ (GMT) Type of image: _______

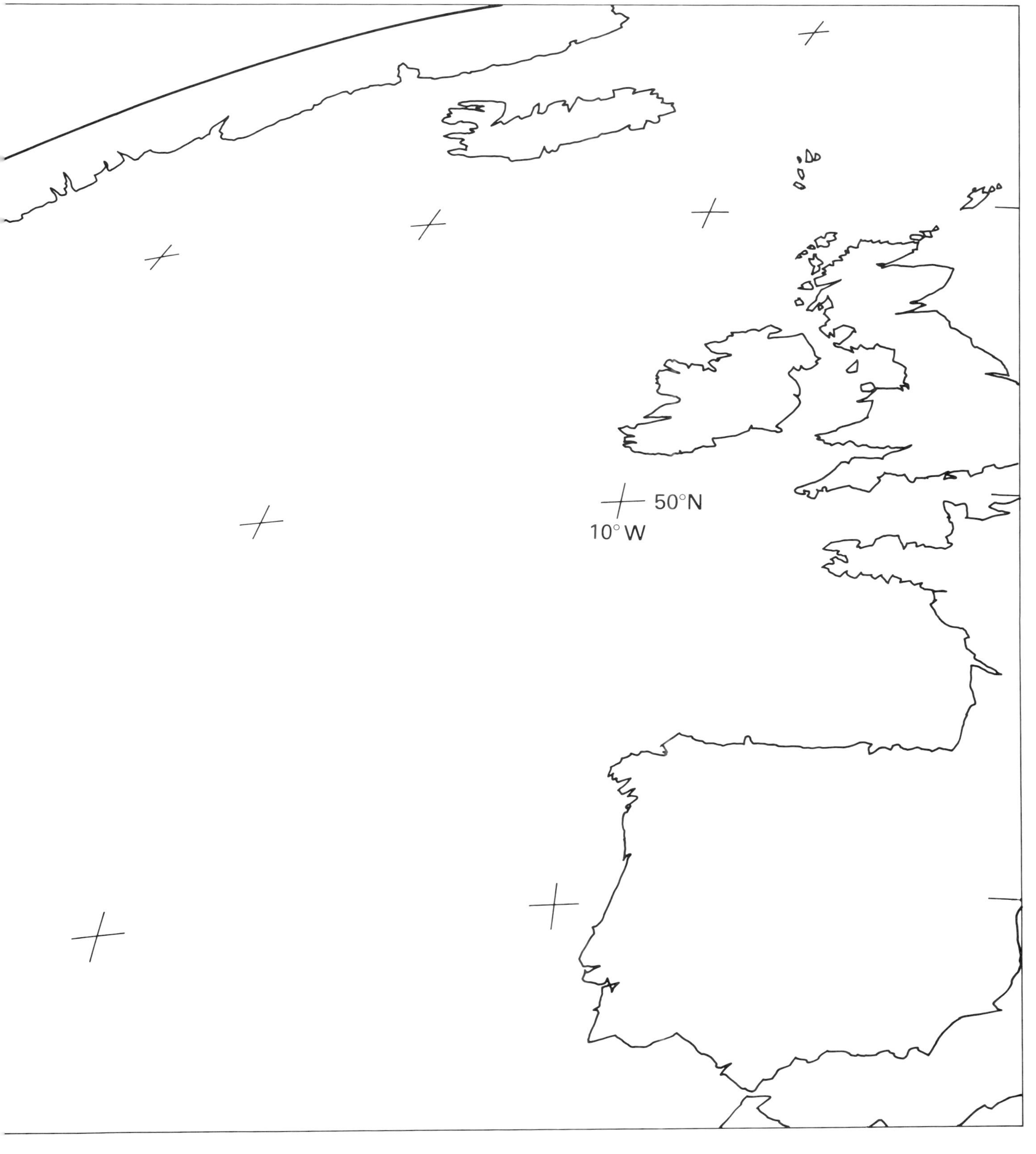

Meteosat base map, section C2

Image data **Scale** 1 cm = 100 km

Date: ______________ Time: _______ (GMT) Type of image: _______

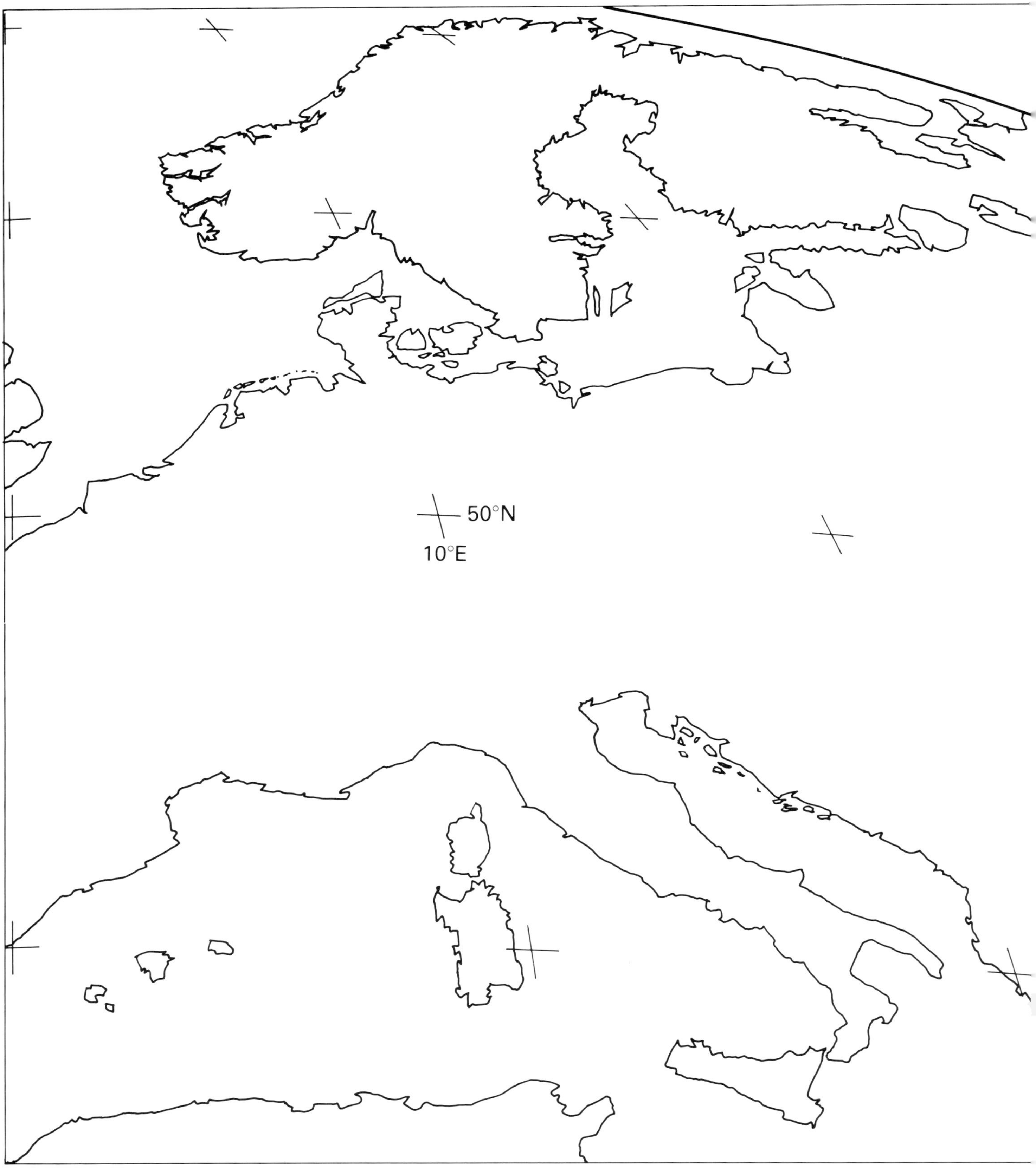

Meteosat base map, section C3

Image data **Scale** 1 cm = 225 km

Date: ______________ Time: _______ (GMT) Type of image: _______

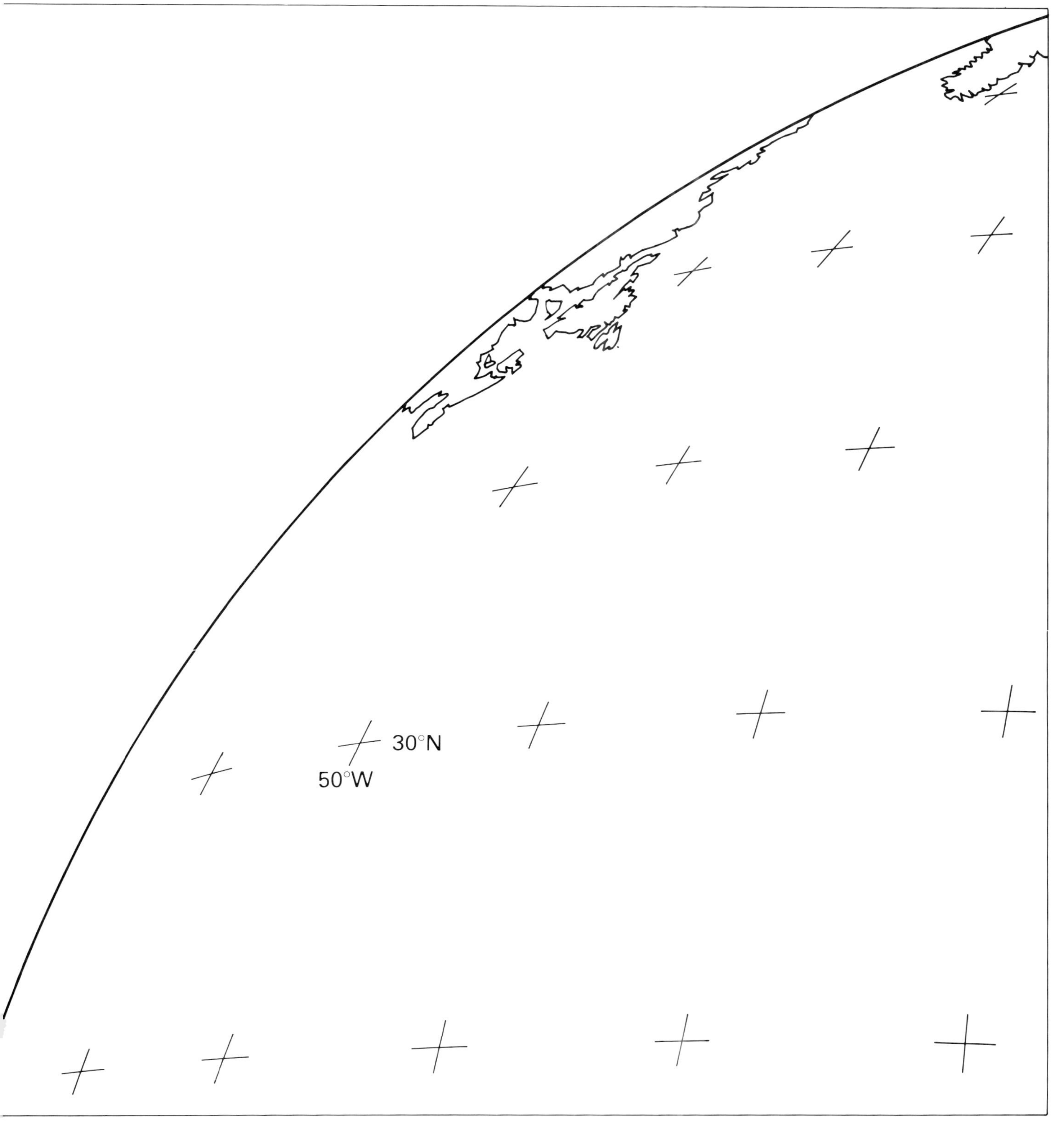

Meteosat base map, section D1

Image data **Scale** 1 cm = 175 km

Date: ____________ Time: ______ (GMT) Type of image: ______

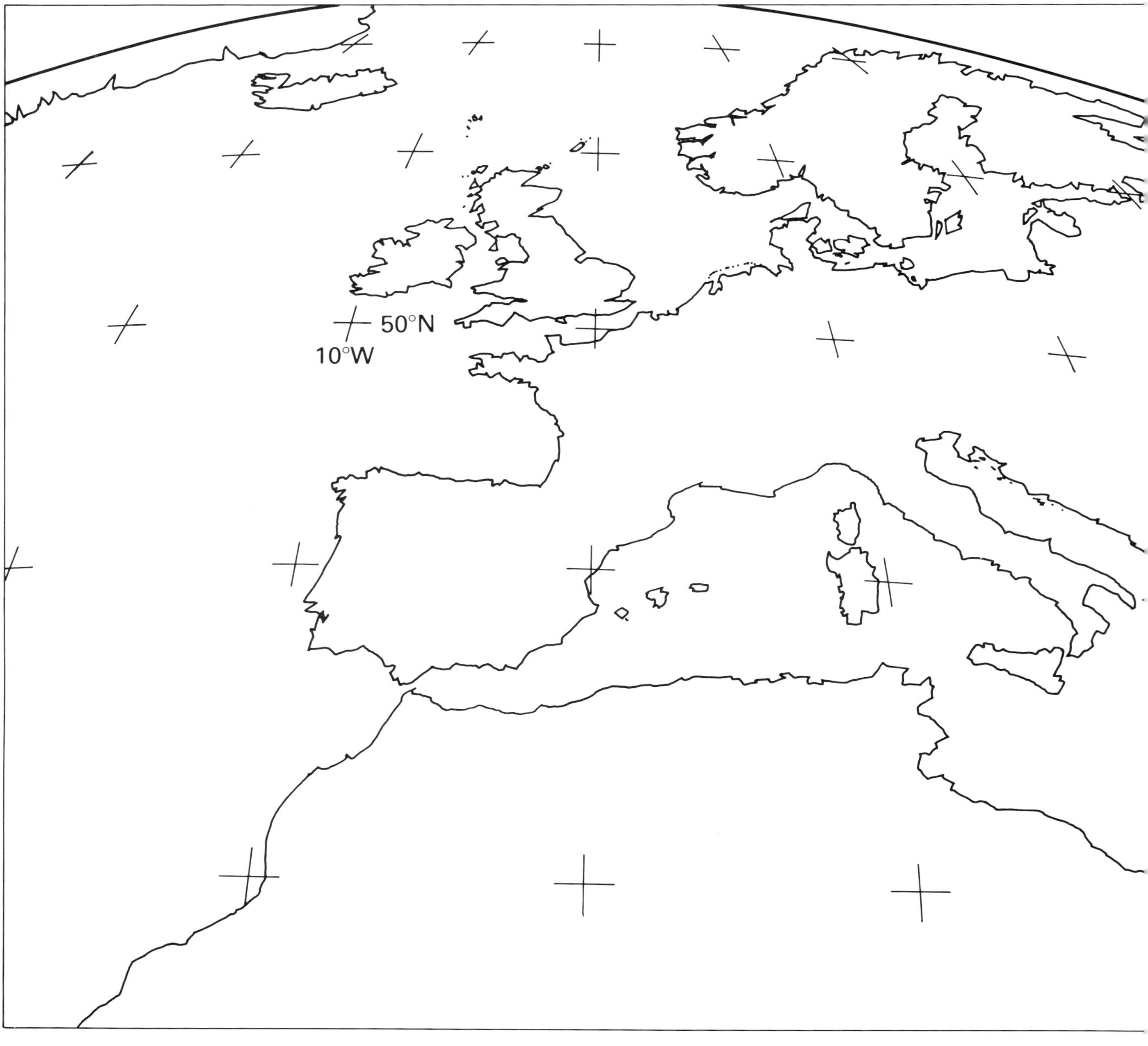

Meteosat base map, section D2

Image data

Scale 1 cm = 225 km

Date: ____________ Time: ______ (GMT) Type of image: ______

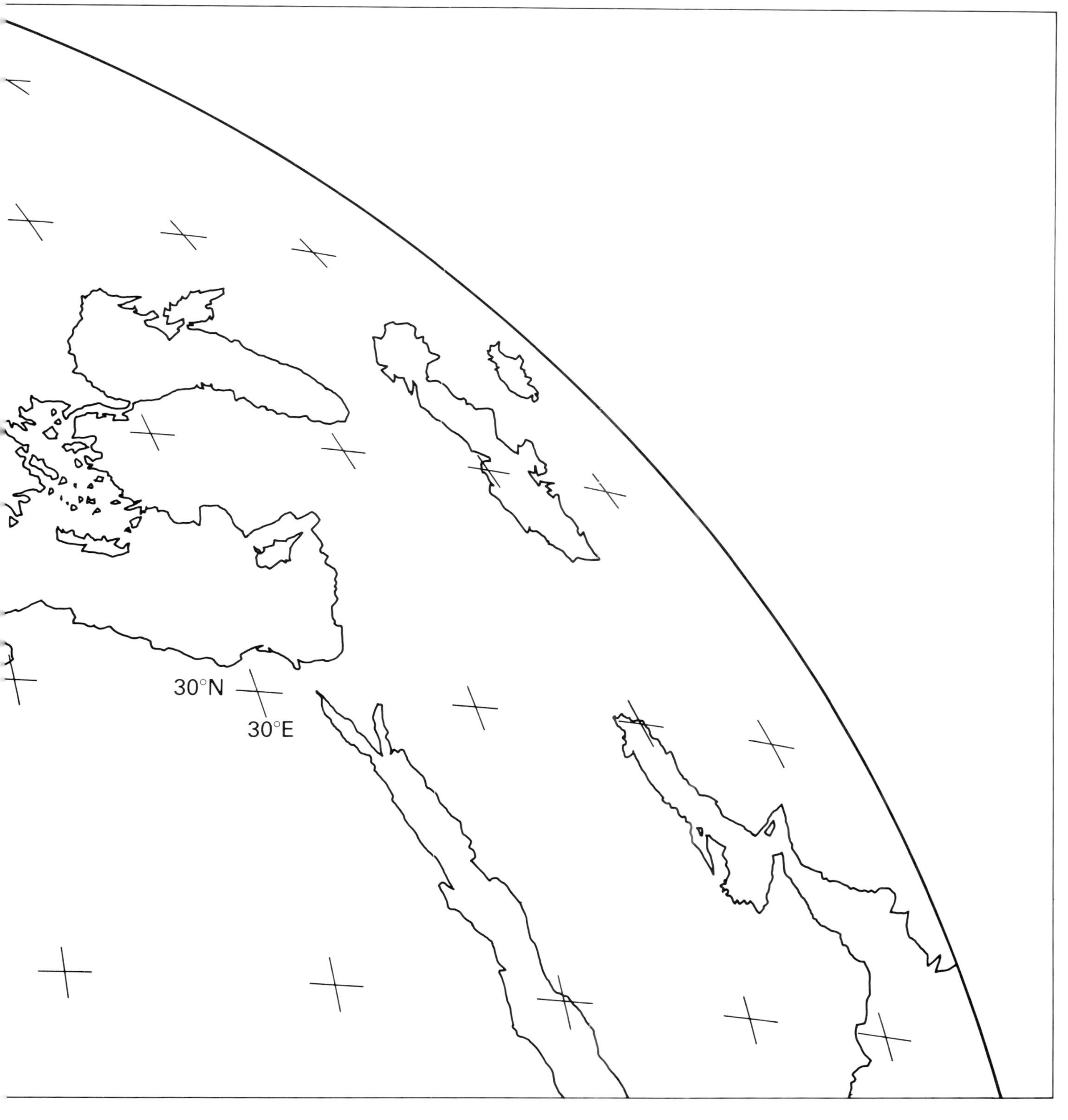

Meteosat base map, section D3

Image data

Scale 1 cm = 320 km

Date: ______________ Time: ________ (GMT) Type of image: ________

Meteosat base map, section D4

Image data

Scale 1 cm = 180 km

Date: ____________ Time: ______ (GMT) Type of image: ______

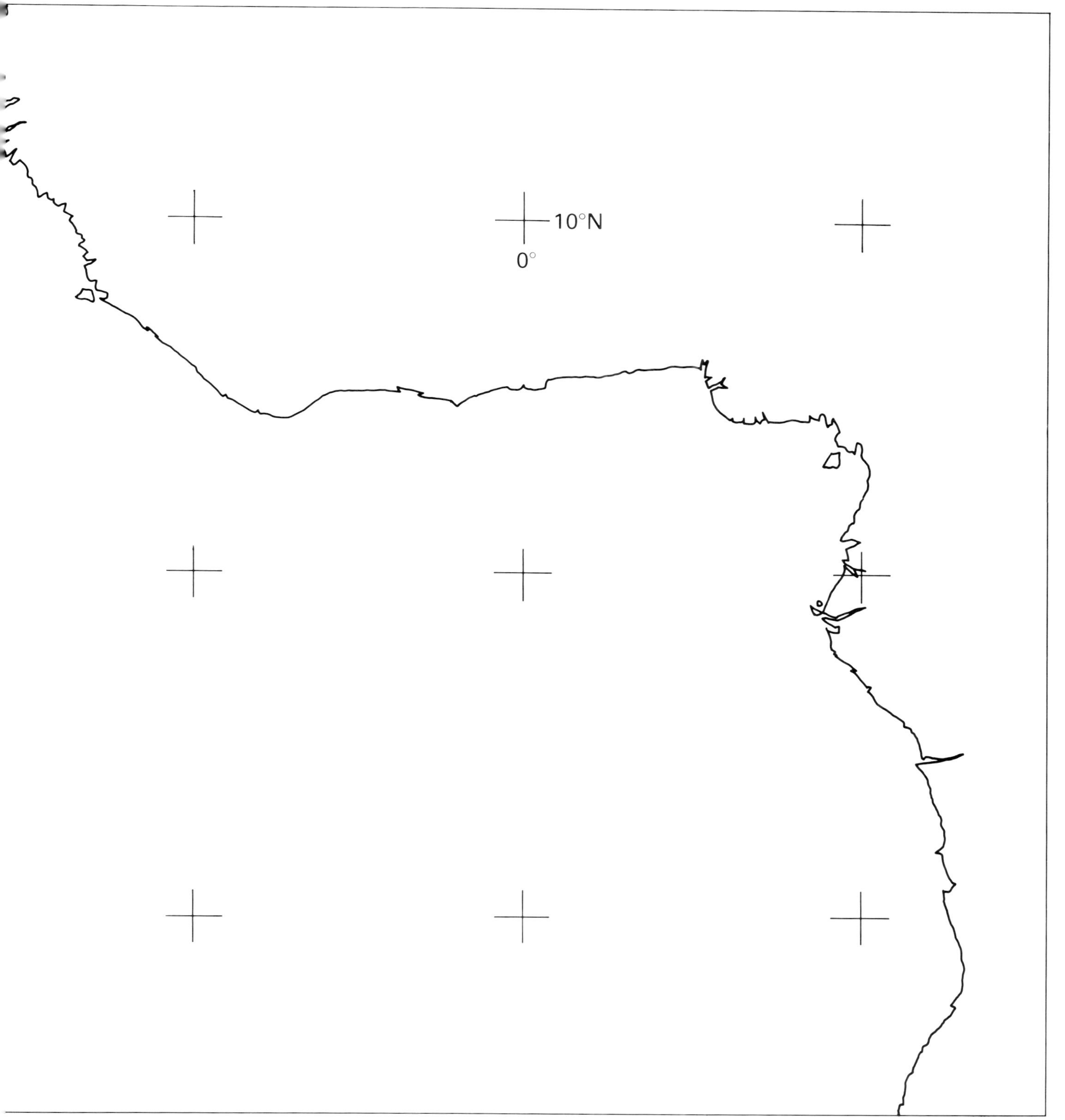

Meteosat base map, section D5

Image data

Scale 1 cm = 212 km

Date: ______________ Time: _______ (GMT) Type of image: _______

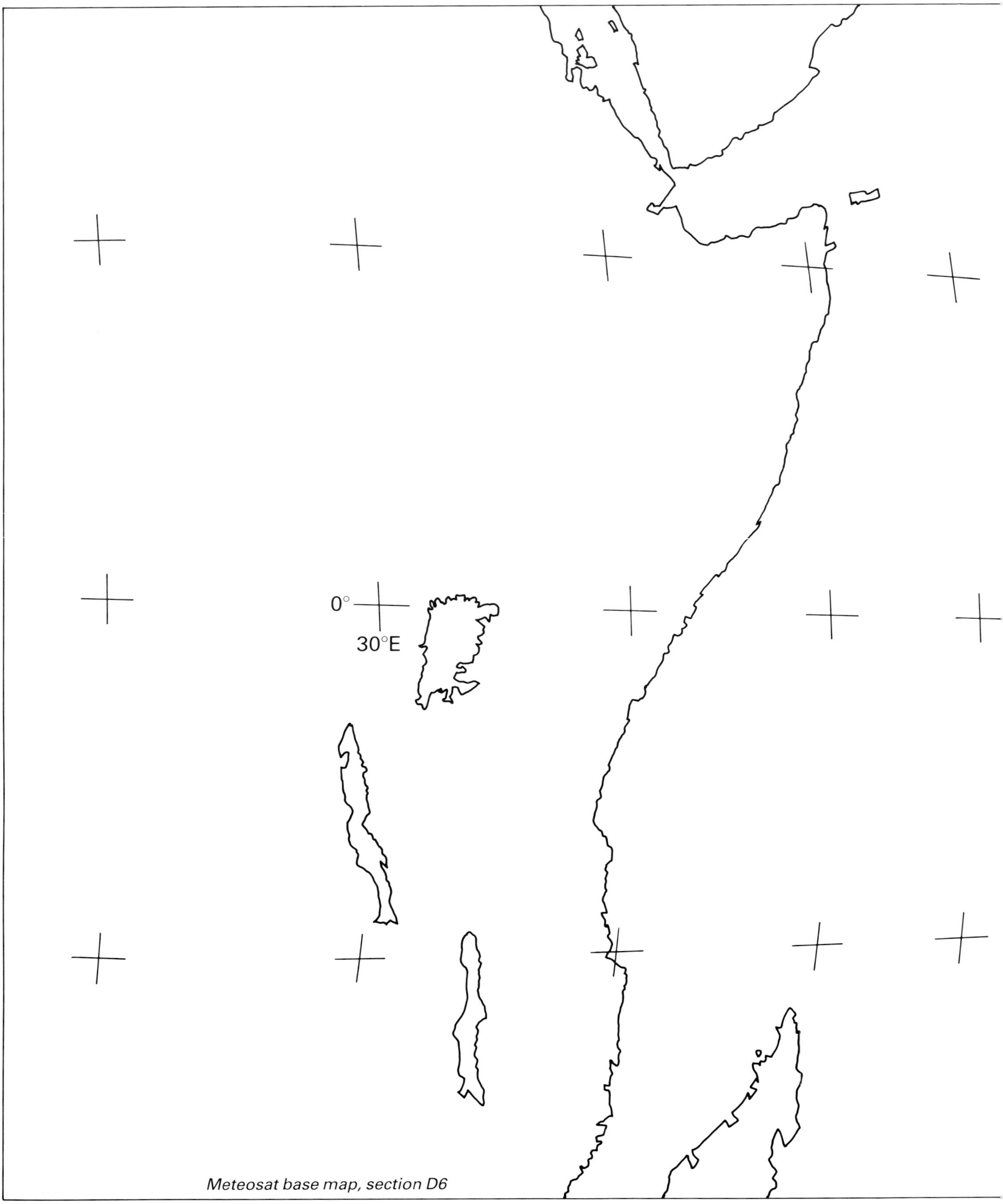

Meteosat base map, section D6

Image data

Scale 1 cm = 275 km

Date: ______________ Time: ________ (GMT) Type of image: ________

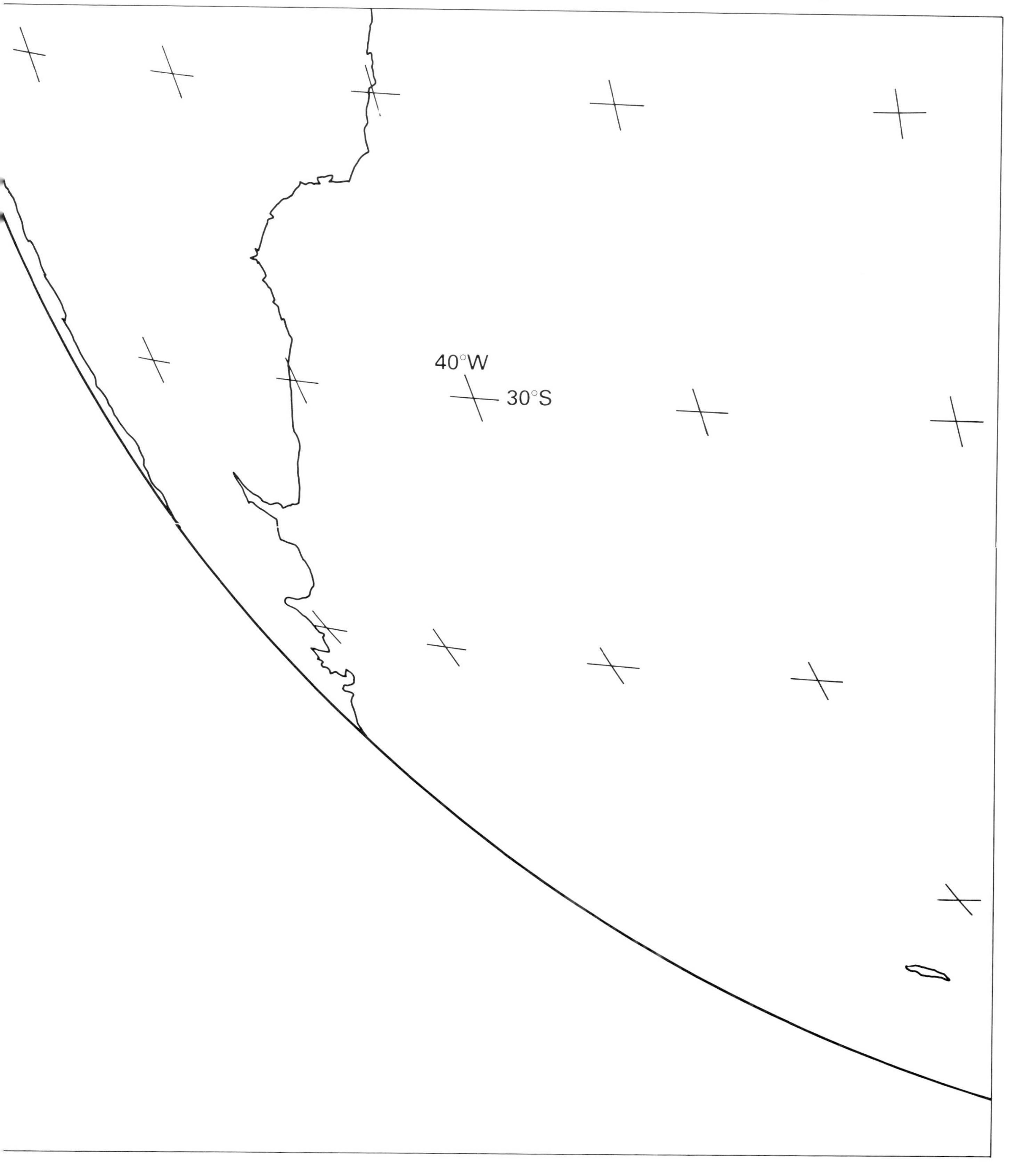

Meteosat base map, section D7

Image data **Scale** 1 cm = 175 km

Date: ______________ Time: _______ (GMT) Type of image: _______

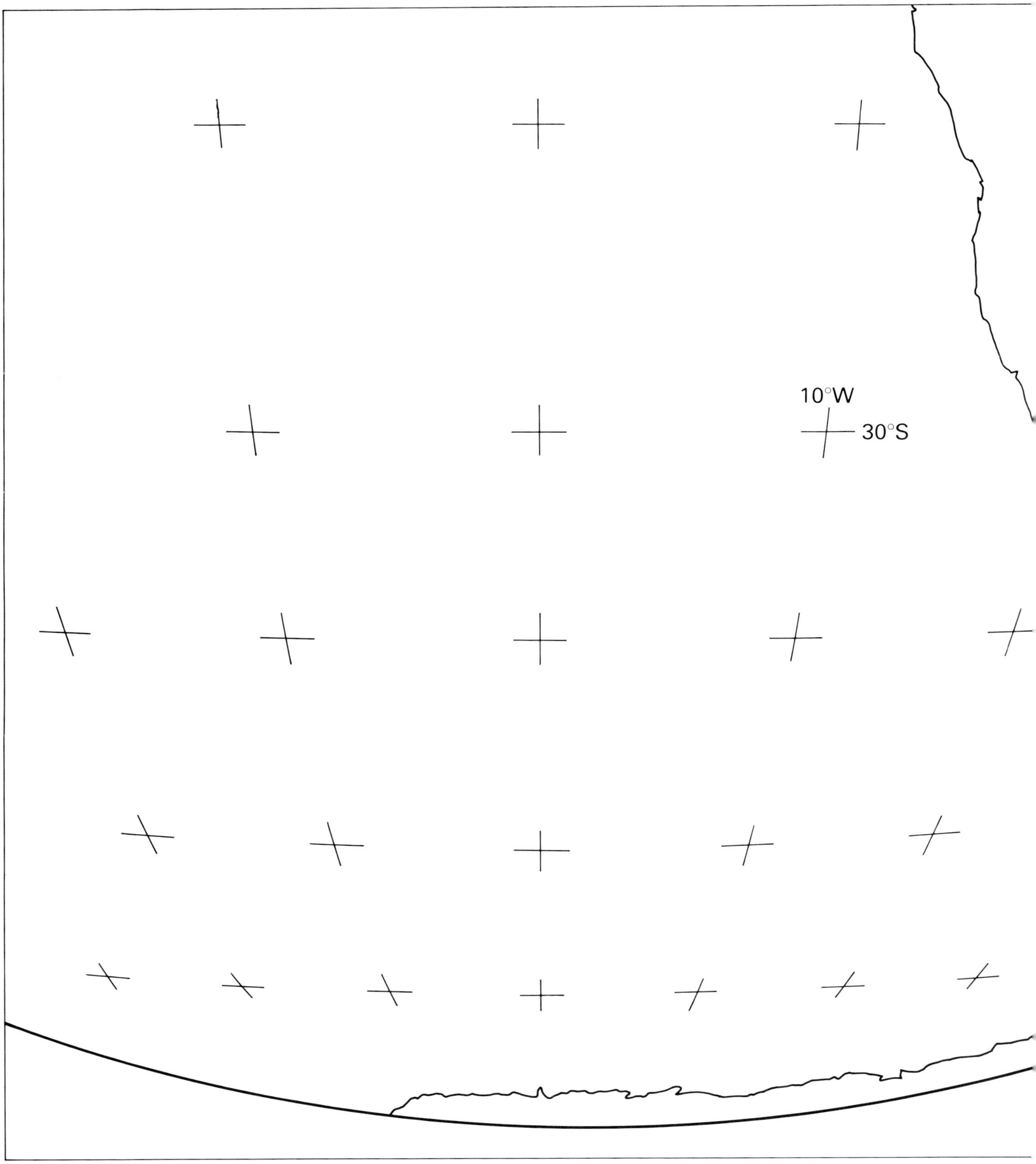

Meteosat base map, section D8

Image data

Scale 1 cm = 230 km

Date: ______________ Time: _______ (GMT) Type of image: _______

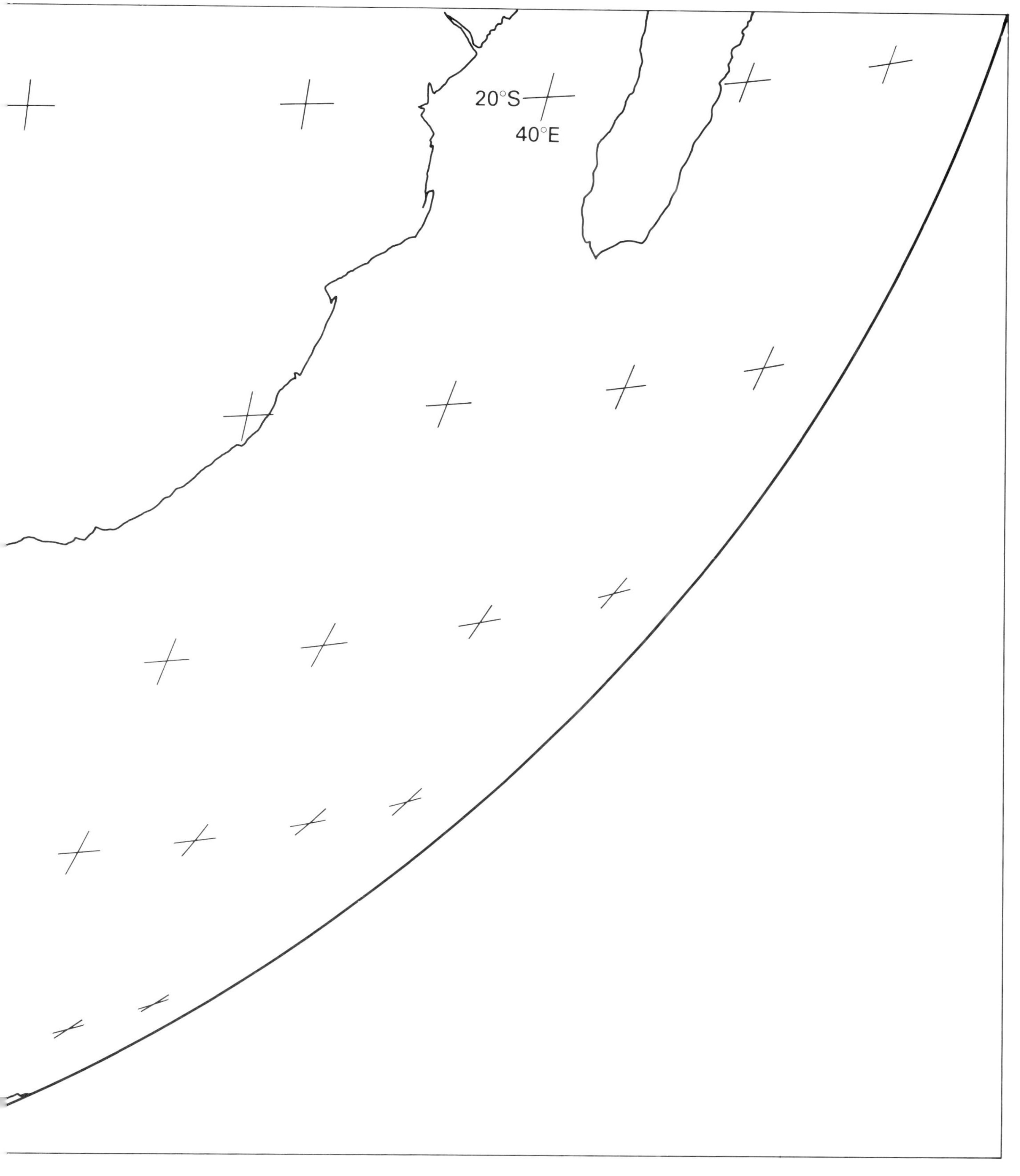

Meteosat base map, section D9

Image data **Scale** (at the equator only) 1 cm = 620 km

Date: ______________ Time: ________ (GMT) Type of image: ________

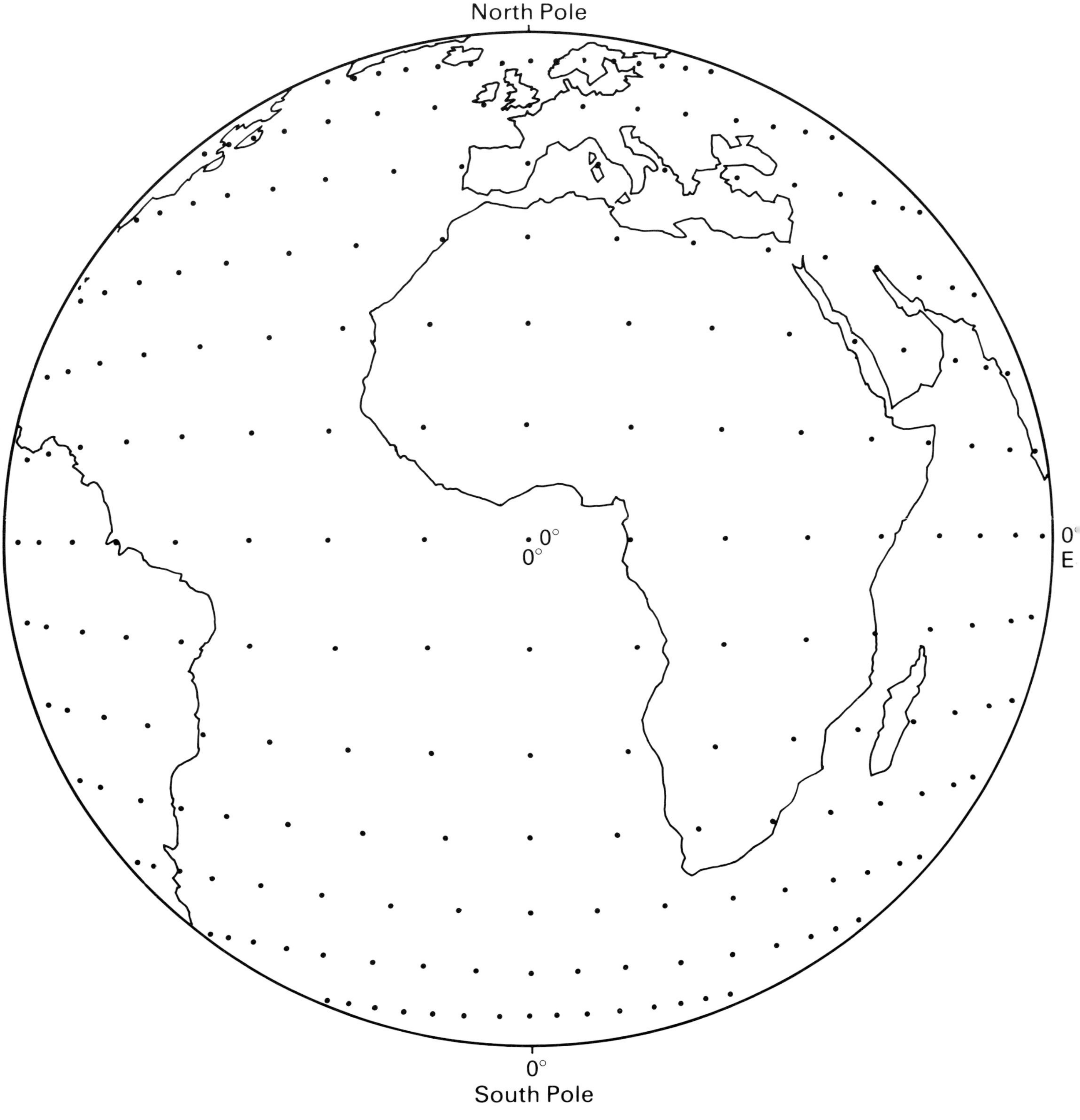

Meteosat base map, whole earth

Image data **Scale** 1 cm = 250 km

Date: ______________ Time: _______ (GMT) Type of image: _______

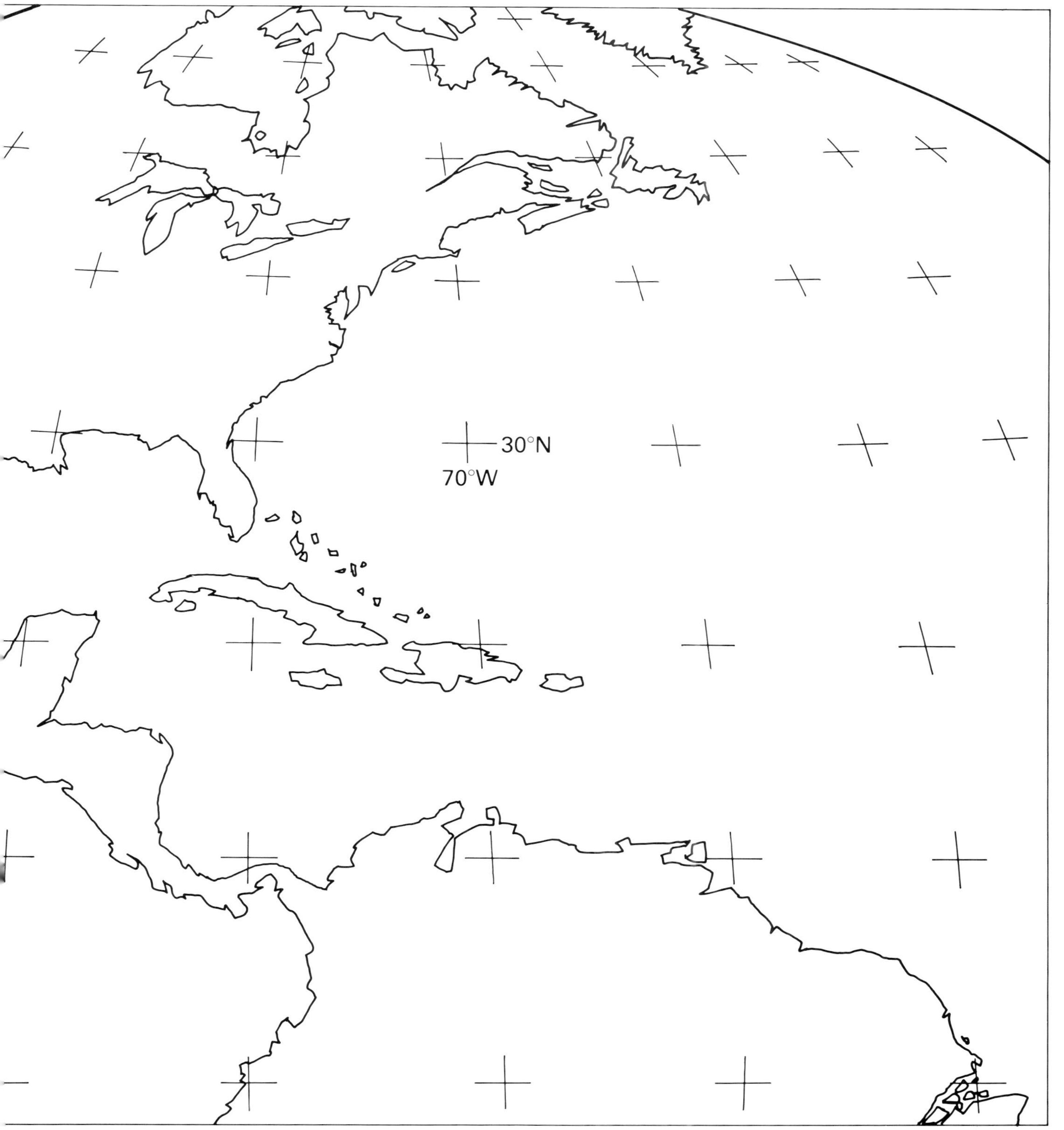

GOES East base map, North America

Image data **Scale** 1 cm = 230 km

Date: ____________ Time: ______ (GMT) Type of image: ______

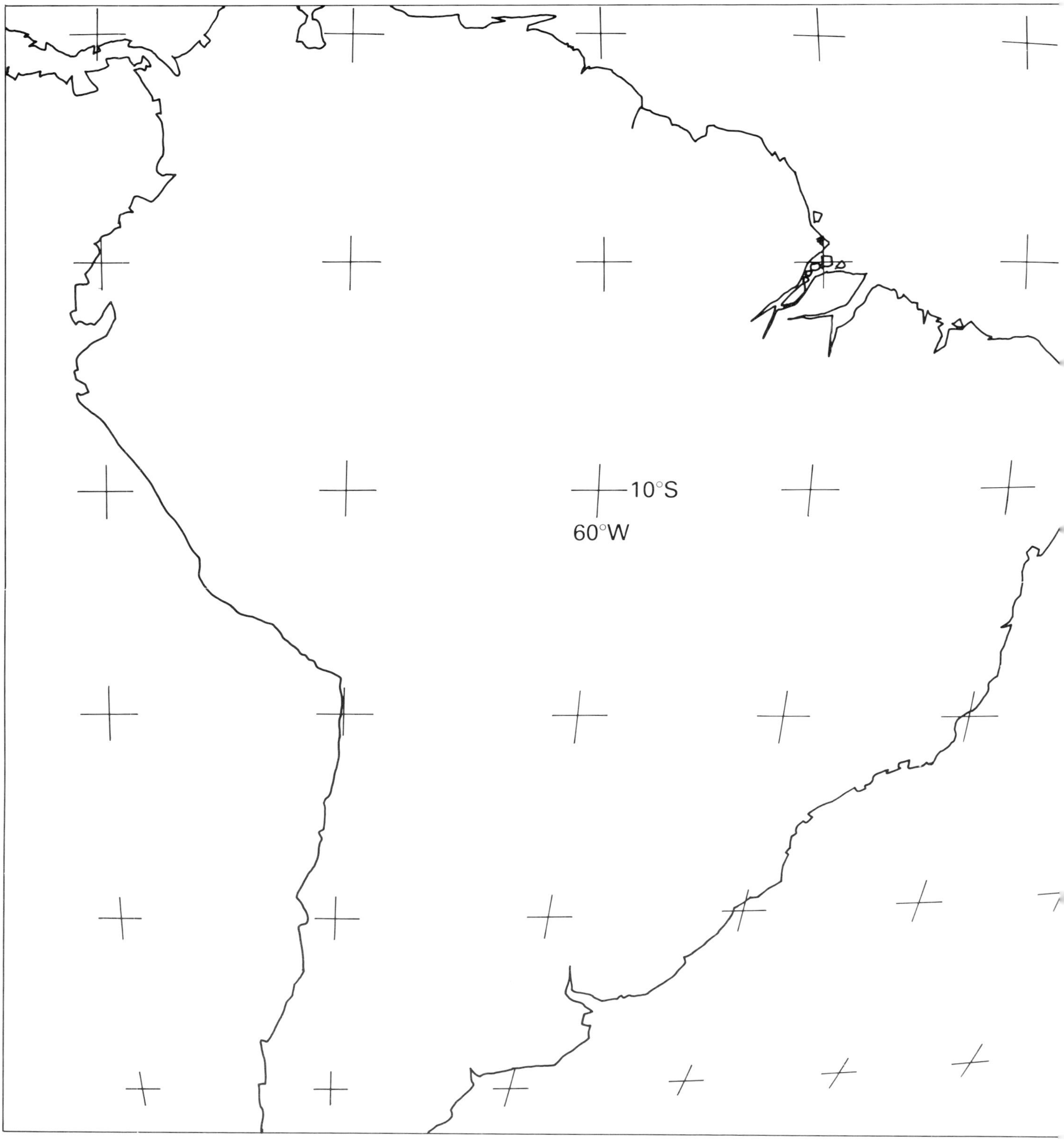

GOES East base map, South America

Radio

Local radio	Produces accurate local forecasts at different times of the day.
Radios 1 and 2	Produce only very short forecasts.
Radio 3	At 06.55, gives a 4-minute forecast for the UK and inshore waters.
Radio 4	Gives detailed forecasts at: 06.55, 07.55, 12.55 and 17.55. Gives the shipping forecasts at: 05.55, 13.55, 17.50 and 00.30. Other forecasts can be heard at: 06.02 and 00.10. On *Farming Today* there is a long-range forecast. This can be heard at about 06.20 each day Monday to Saturday, but not on Sunday.

Television

BBC1	The most important forecasts are at the following times: 07.50, 08.50, 09.03, 11.03, 13.30, 18.30, 21.25 and 23.00. (These vary over a weekend.) On most occasions there are details of the weather over Europe. At 11.03 there are details about European weather and also at 21.25 on Fridays and 23.00 on Thursdays. There is a 12.58 forecast on Sundays which gives a long-range weekly outlook.
BBC2, ITV and Channel 4 (S4C)	These channels give much less detailed forecasts.

Newspapers

The Times *Daily Telegraph* *The Guardian* *Independent*	All these papers produce a detailed weather forecast each day, including a weather map for the north Atlantic and often a detailed chart for the UK. They also provide some statistical information about temperature and rainfall in different world locations.
Other papers	Produce varying levels of UK and regional summaries.

Prestel and Ceefax

Weather information is available on each of these services:

Prestel – pages 115 and 209

Ceefax – page 151 and following pages

Telephone

The information on the maps on pages 128–9 gives details of two major telephone weather forecast services.

DETAILED WEATHER INFORMATION FOR EVERY REGION OF THE COUNTRY

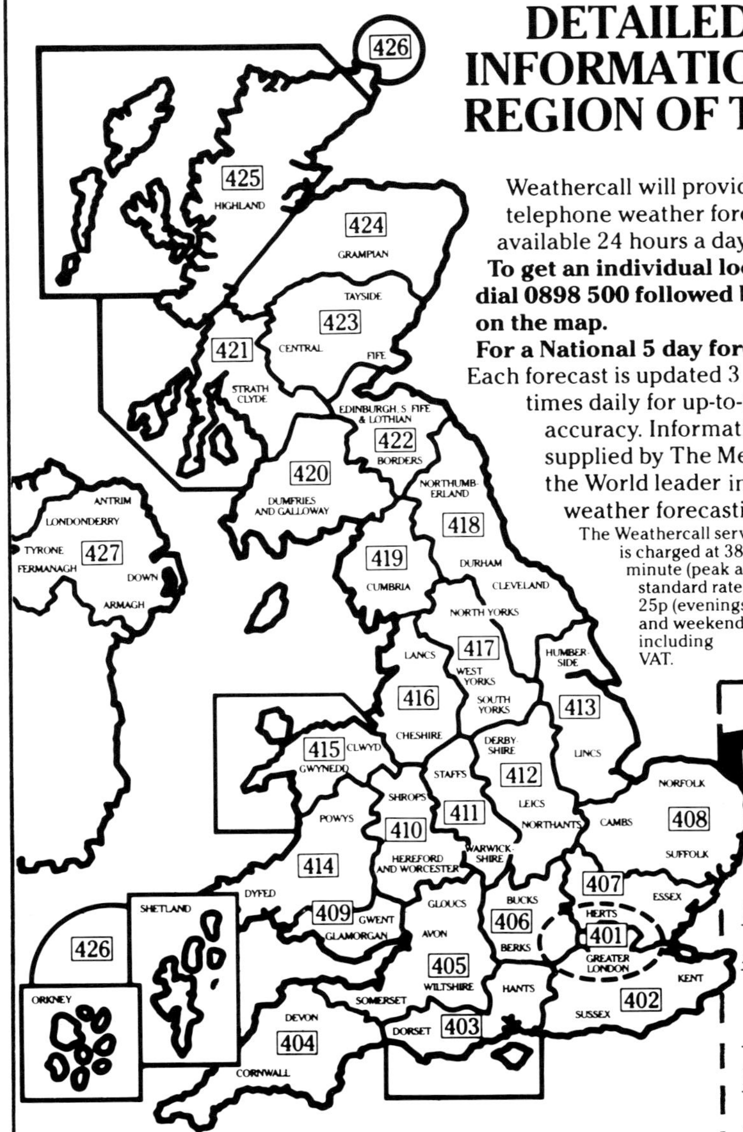

Weathercall will provide you with the most accurate telephone weather forecasting service in Britain, available 24 hours a day, 7 days a week.

To get an individual local forecast for your region, dial 0898 500 followed by the 3 digit code indicated on the map.

For a National 5 day forecast call 0898 500 430.

Each forecast is updated 3 or more times daily for up-to-the minute accuracy. Information is supplied by The Met Office, the World leader in weather forecasting.

The Weathercall service is charged at 38p per minute (peak and standard rate) 25p (evenings and weekends), including VAT.

FREE!

For your free Weathercall card send this coupon to the address below.

Name: ____________________

Address: ____________________

____

____

Postcode: ____________________

Weathercall Dept,
Telephone Information Services Ltd, Dewhurst House,
24 West Smithfield,
London EC1A 9DL.

REMEMBER CALL 0898 500 FOLLOWED BY YOUR LOCAL NUMBER

All Weathercall information is provided exclusively by the Met. Office.

The Met. Office

British Telecom's NEW improved weather service is now more local

PHONE NOW FOR YOUR FREE INDEX CARD OF LOCAL NUMBERS AND DETAILS OF HOW TO WIN A SPECIAL PRIZE

0898 14 12 00

Dial 0898 14 12 .. plus the map number for your area e.g. LONDON...14

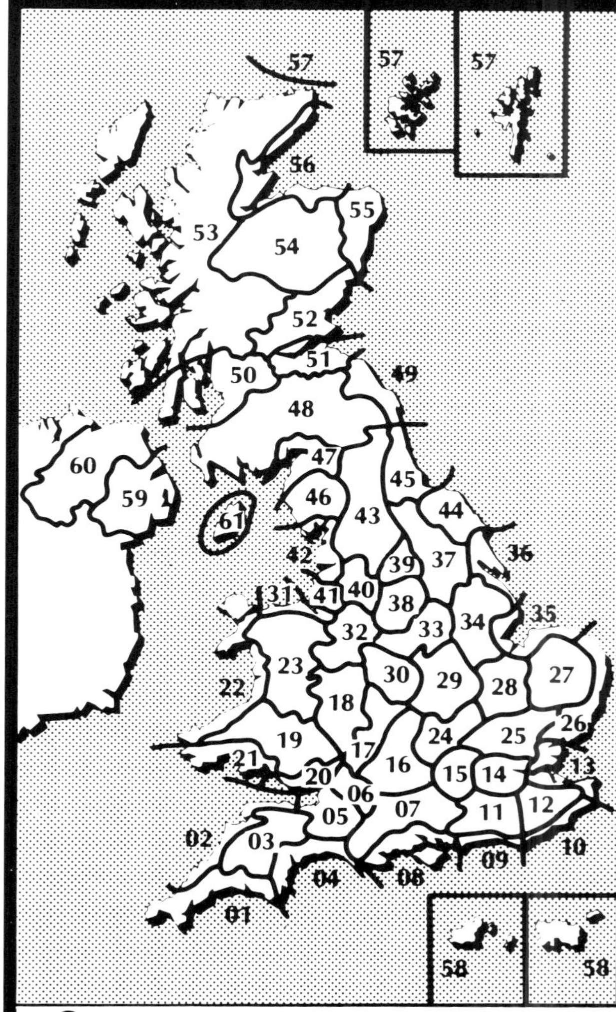

NATIONAL OUTLOOK

70 National 5 day picture.

5 DAY FORECASTS

71 SCOTLAND and NORTHERN IRELAND
72 WALES
73 Northern ENGLAND
74 Midlands
75 South West ENGLAND
76 South East ENGLAND

TRAVEL

81 South East ENGLAND
82 Midlands
83 Northern ENGLAND
84 WALES and South West ENGLAND

HOLIDAY RESORTS

91 North America
92 Mainland Spain
93 Balearics
94 Mainland Greece
95 Greek Islands
96 Portugal

OTHER

89 Sporting and Outdoor event weather

PERMANENT NUMBERS ✂ CUT OUT AND KEEP

MORE ACCURATE – Experts analyse your area
MORE FREQUENT – Updated 3 times a day

Dial now for latest update

Local WEATHERDIAL

THE ONLY RELIABLE THING ABOUT THE BRITISH WEATHER

Calls are charged at 38p per minute peak and standard rate, and 25p off peak.

Note that for coastal sea areas, forecasts are normally broadcast in the following order: Viking, North Utsire, South Utsire, Forties, Cromarty, Forth, Tyne, Dogger, Fisher, German Bight, Humber, Thames, Dover, Wight, Portland, Plymouth, Biscay, Trafalgar, Finisterre, Sole, Lundy, Fastnet, Irish Sea, Shannon, Rockall, Malin, Hebrides, Bailey, Fair Isle, Faeroes, South East Iceland.

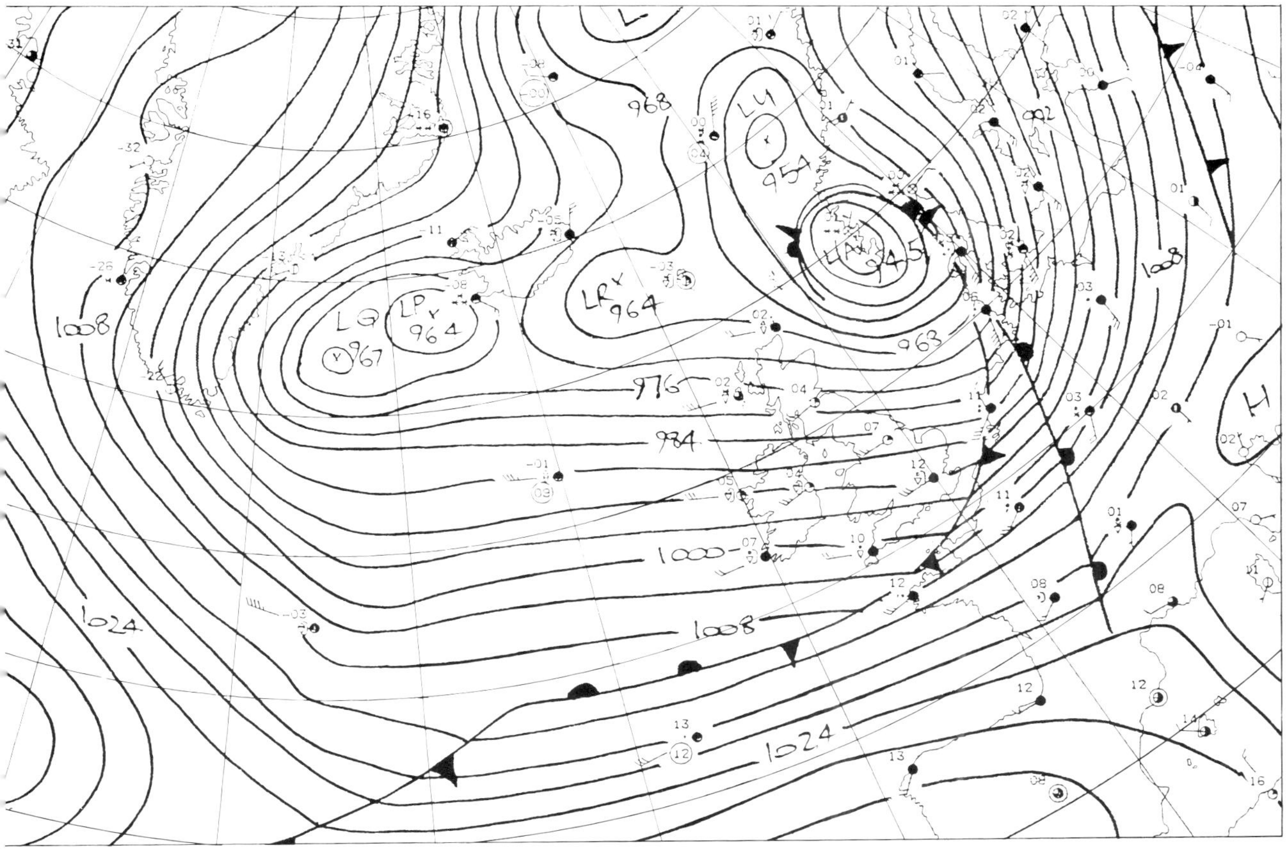

Weather map for 12.00 on 13 January 1984 (Image 5) (Crown copyright reserved)

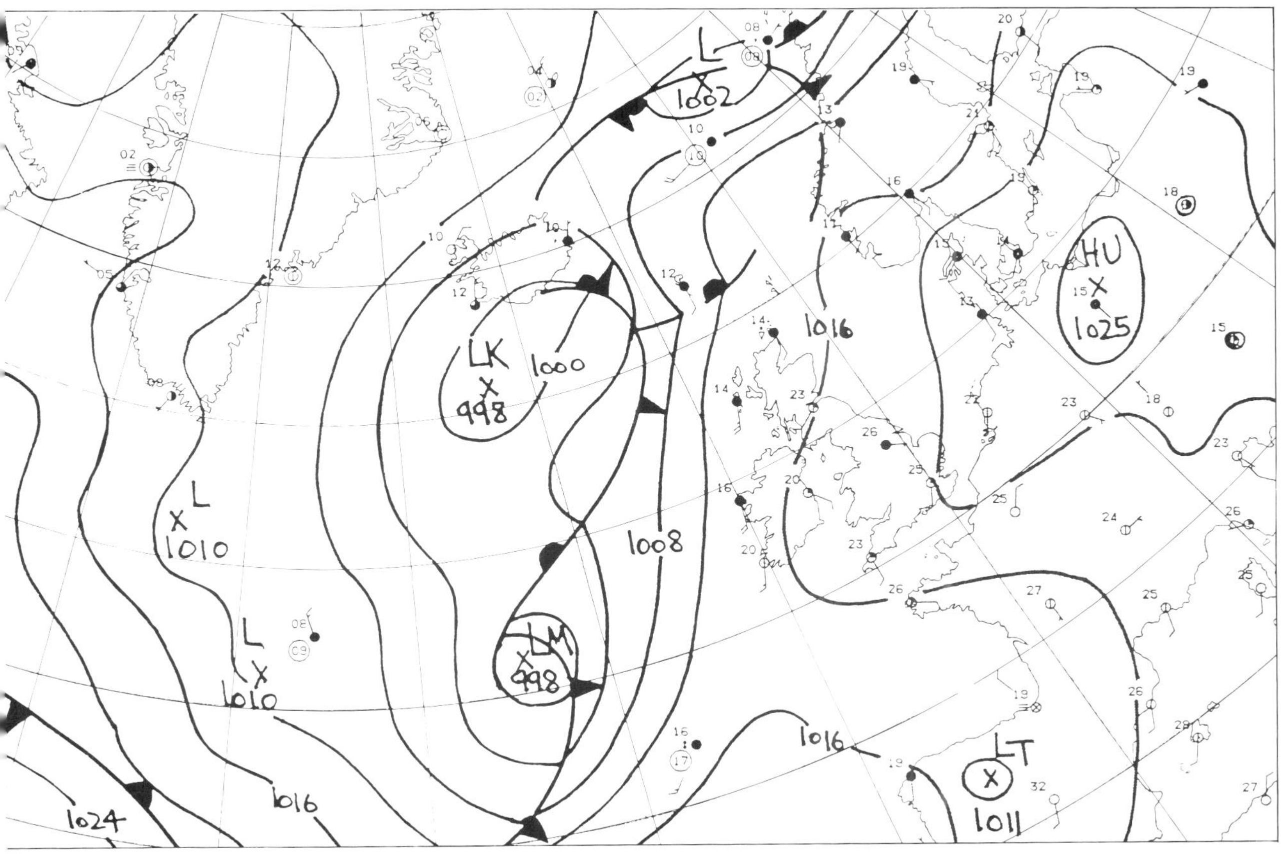

Weather map for 12.00 on 7 July 1984 (Images 1 and 2) (Crown copyright reserved)

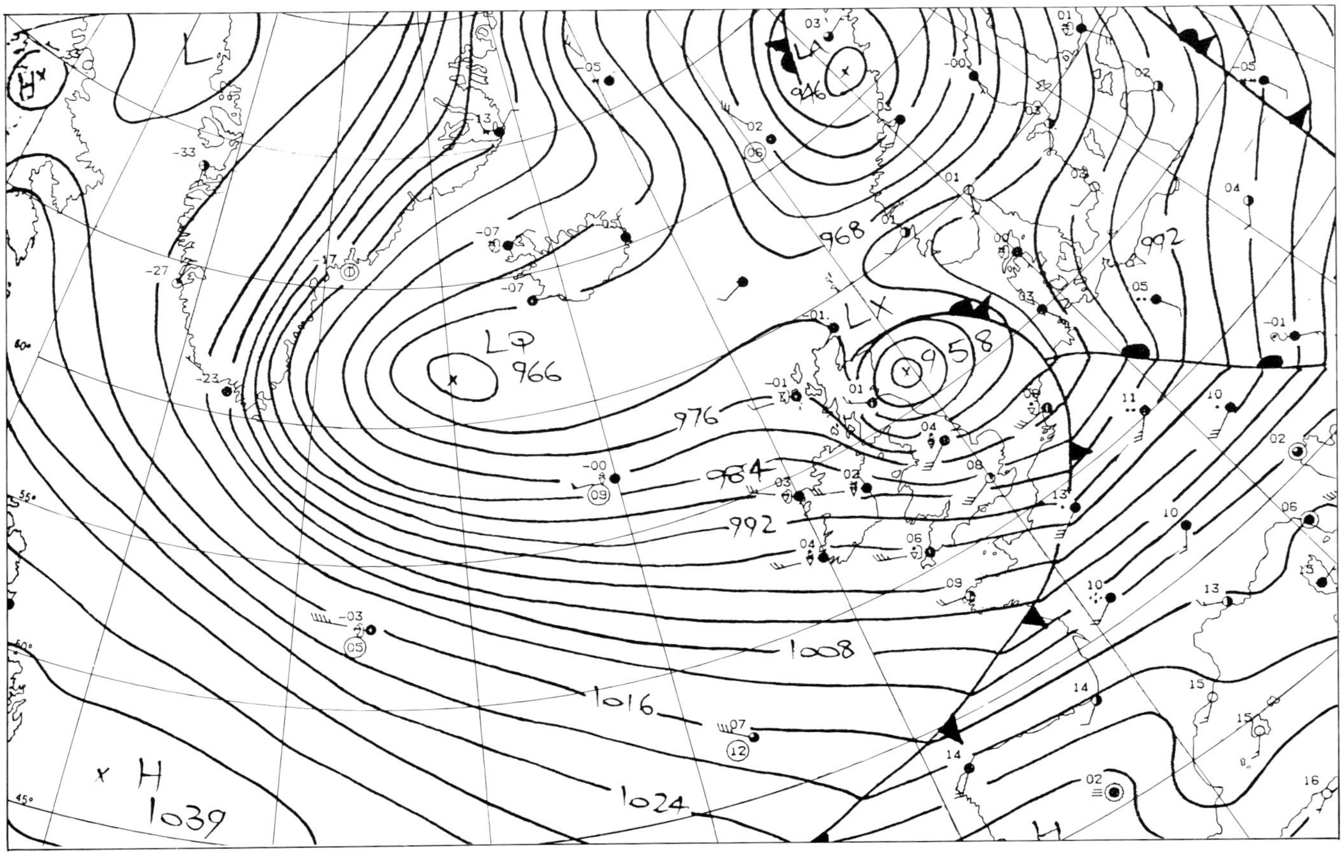

Weather map for 12.00 on 14 January 1984 (Image 7) *(Crown copyright reserved)*

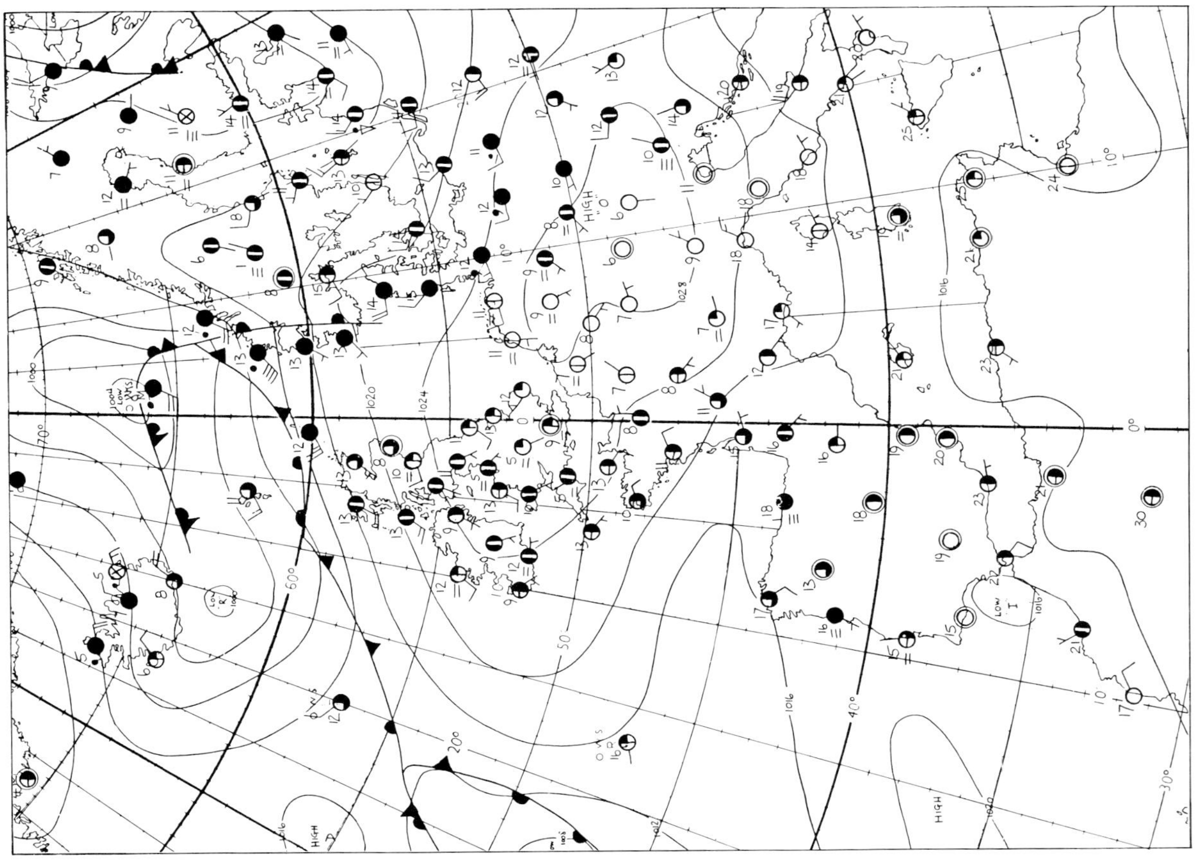

Weather map for 06.00 on 29 August 1979 (Image 6) *(Crown copyright reserved)*

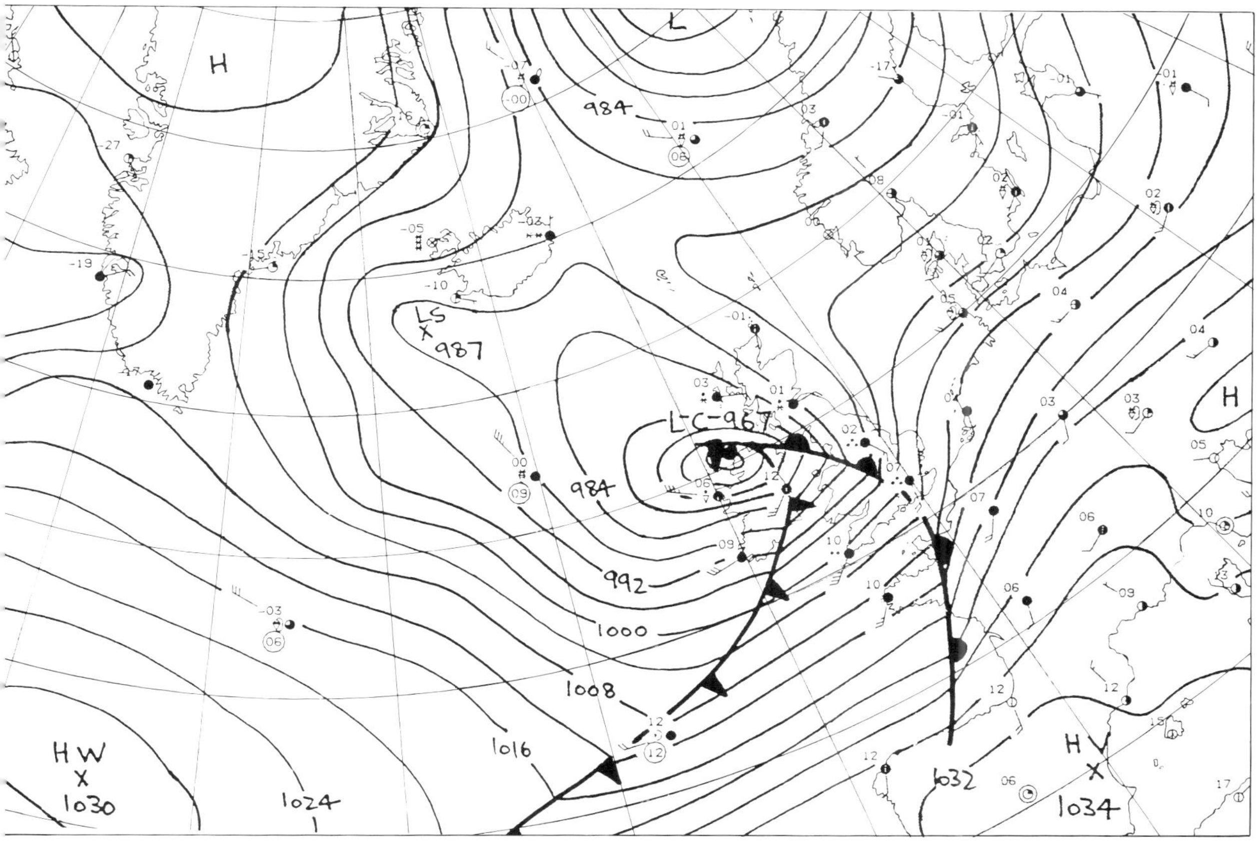

Weather map for 12.00 on 16 January 1984 (Image 9) (Crown copyright reserved)

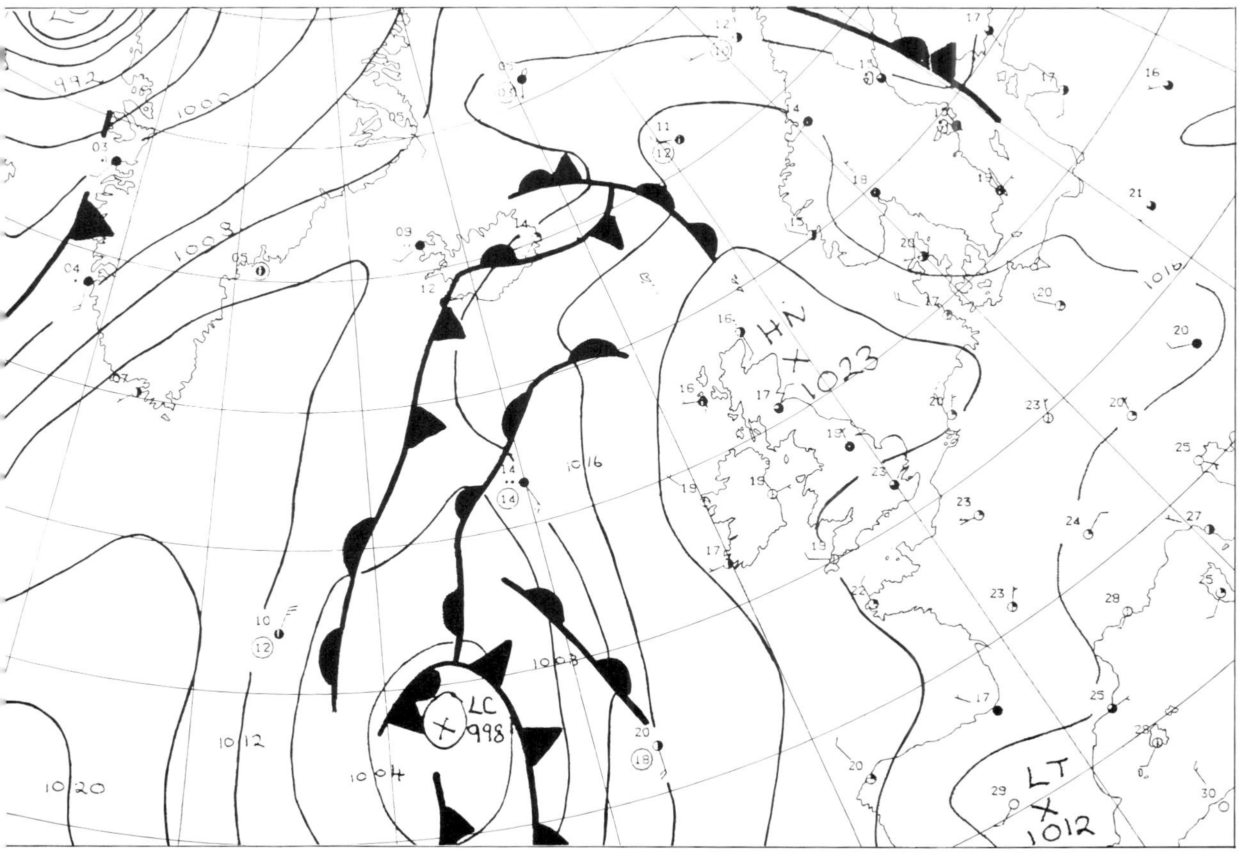

Weather map for 12.00 on 16 August 1984 (Image 8) (Crown copyright reserved)

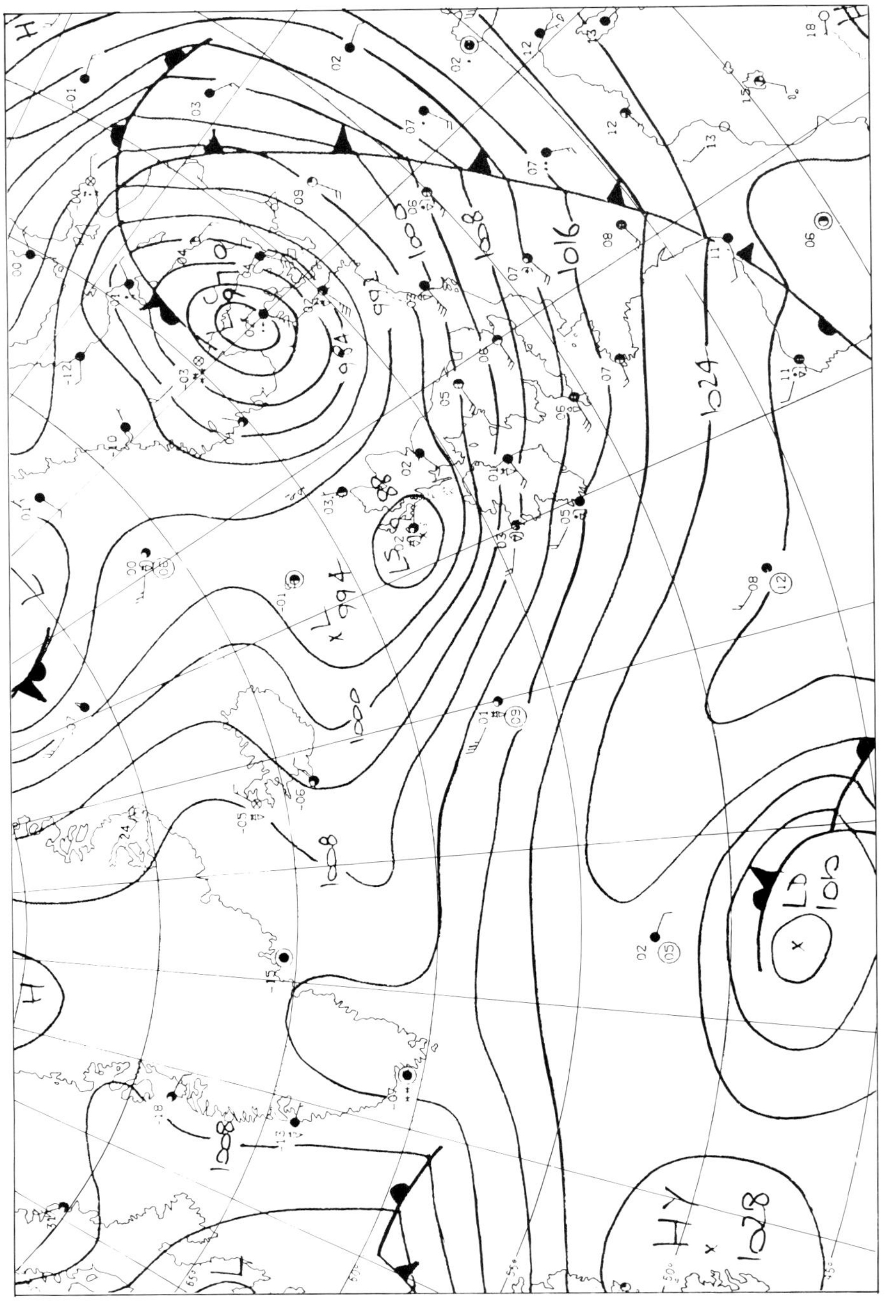

Weather map for 12.00 on 17 January 1984 (Image 10) *(Crown copyright reserved)*

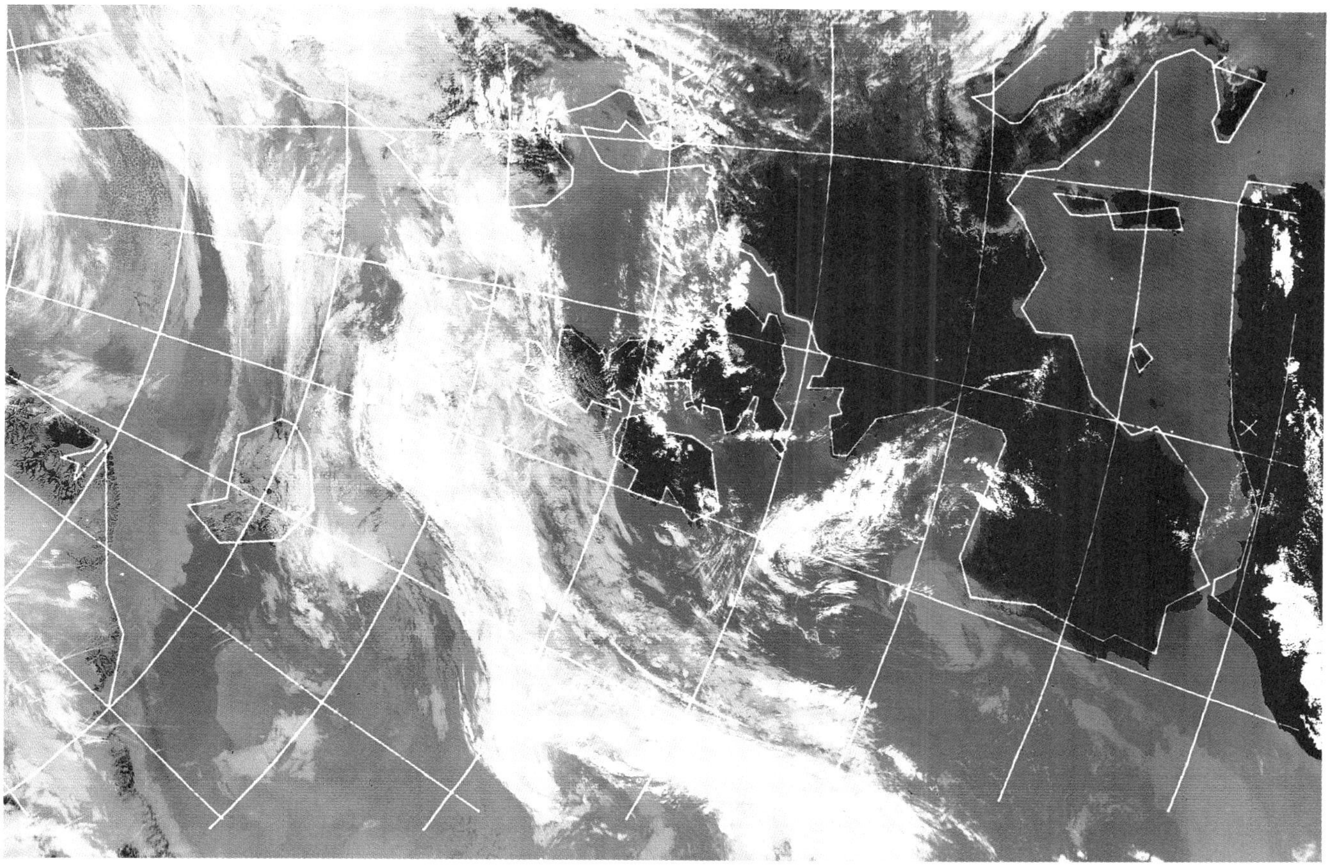

Image 2 – infra-red, 7 July 1984, 15.18 © University of Dundee

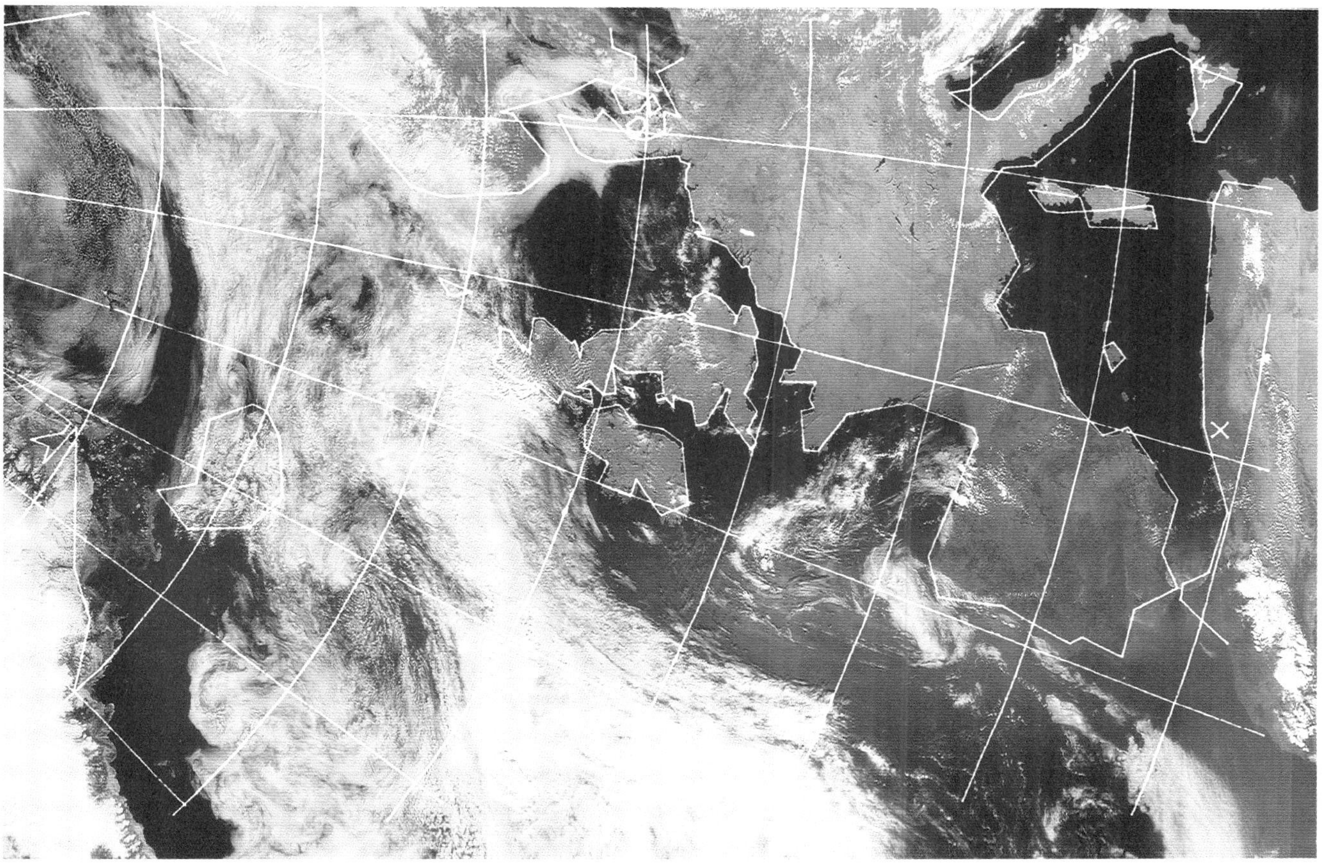

Image 1 – visible, 7 July 1984, 15.18 © University of Dundee

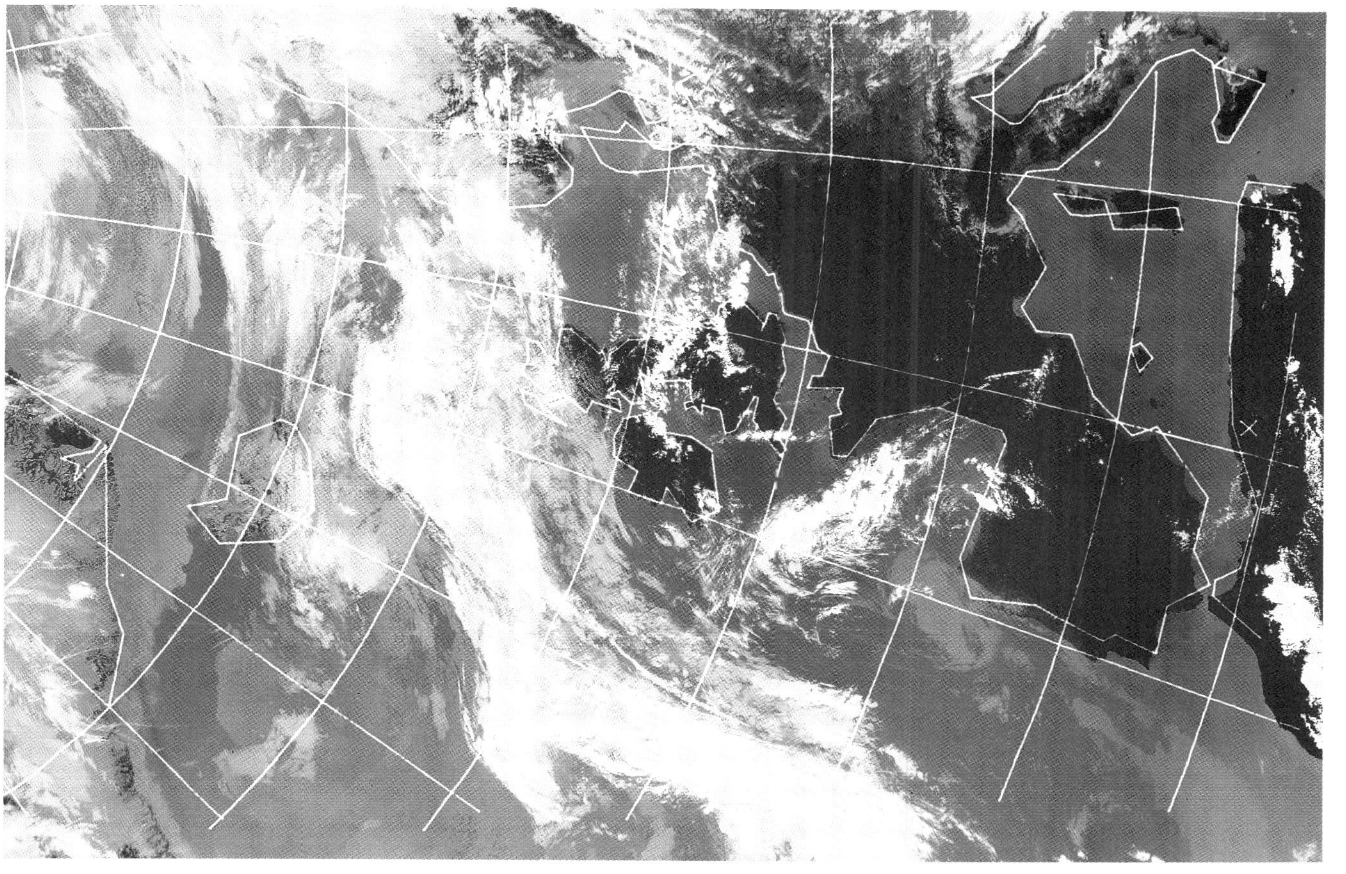

Image 2 – infra-red, 7 July 1984, 15.18 © University of Dundee

Image 1 – visible, 7 July 1984, 15.18 © University of Dundee

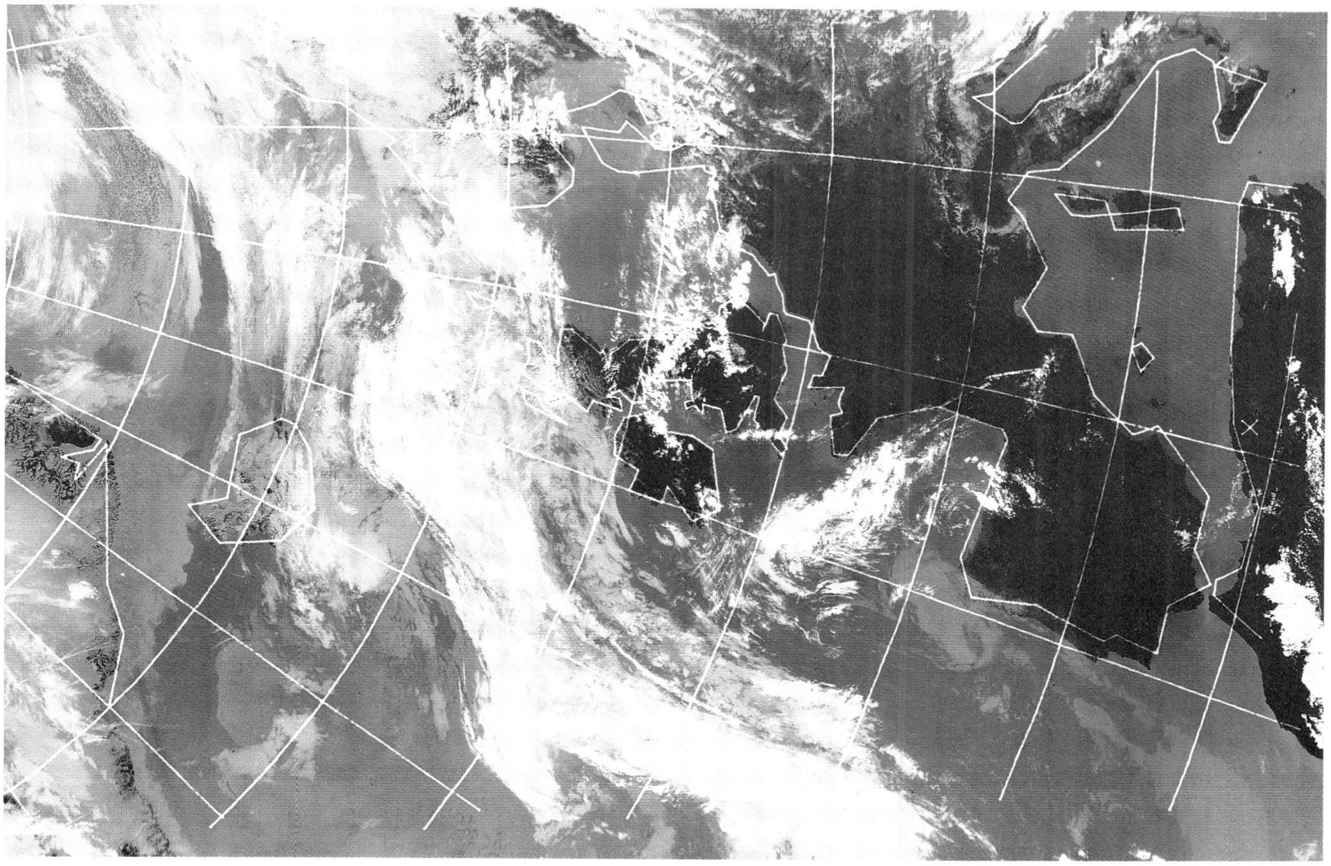

Image 2 – infra-red, 7 July 1984, 15.18 © University of Dundee

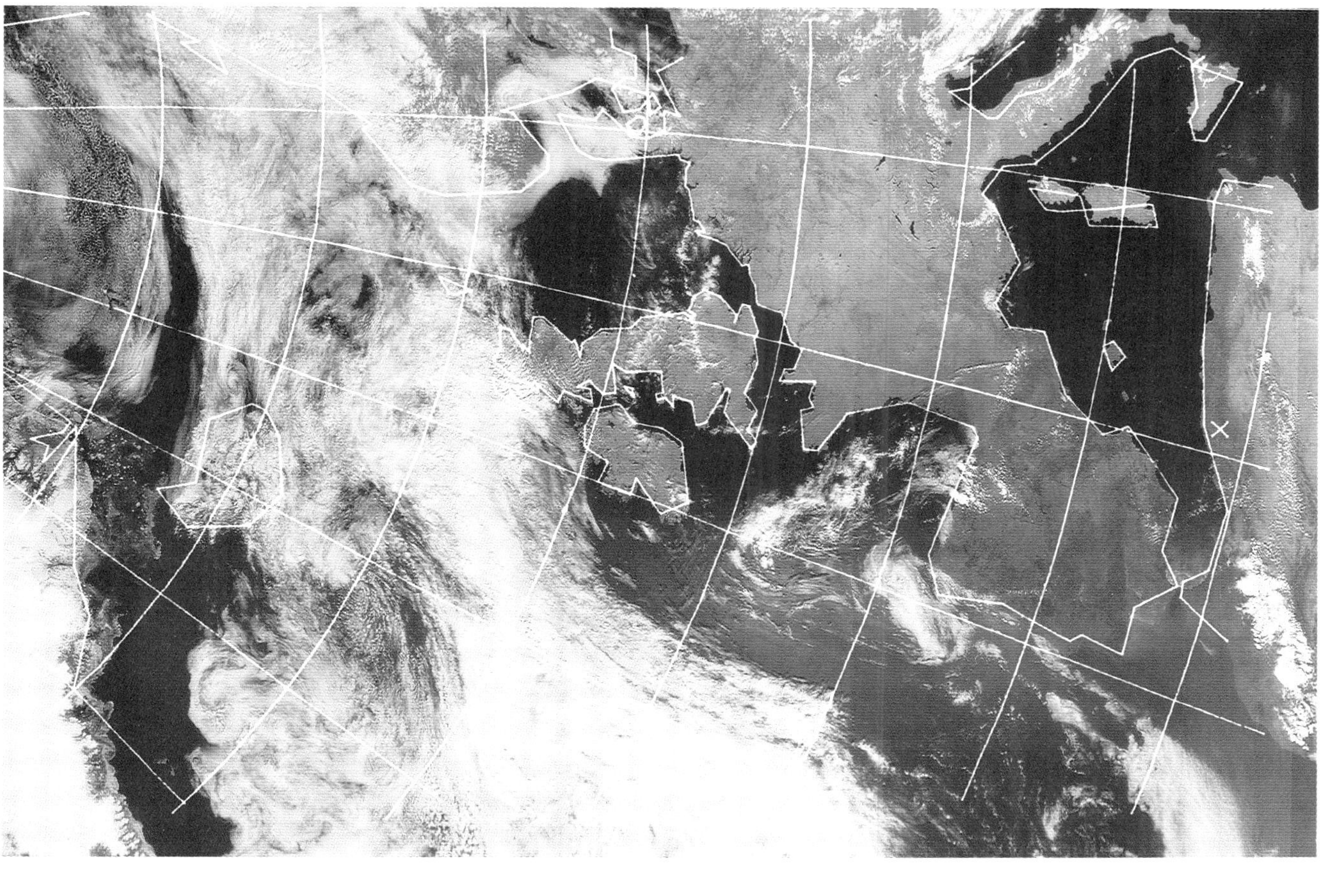

Image 1 – visible, 7 July 1984, 15.18 © University of Dundee

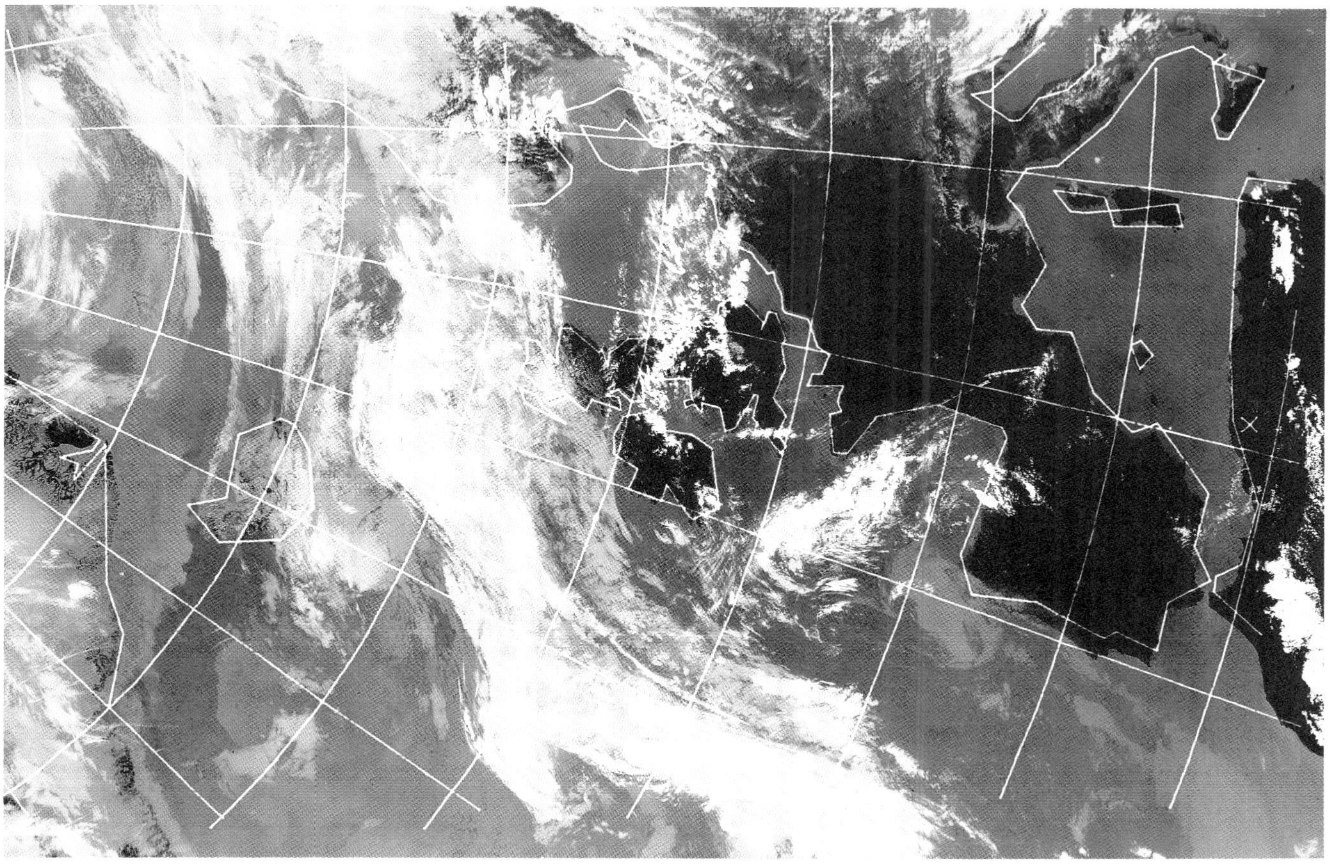

Image 2 – infra-red, 7 July 1984, 15.18 © University of Dundee

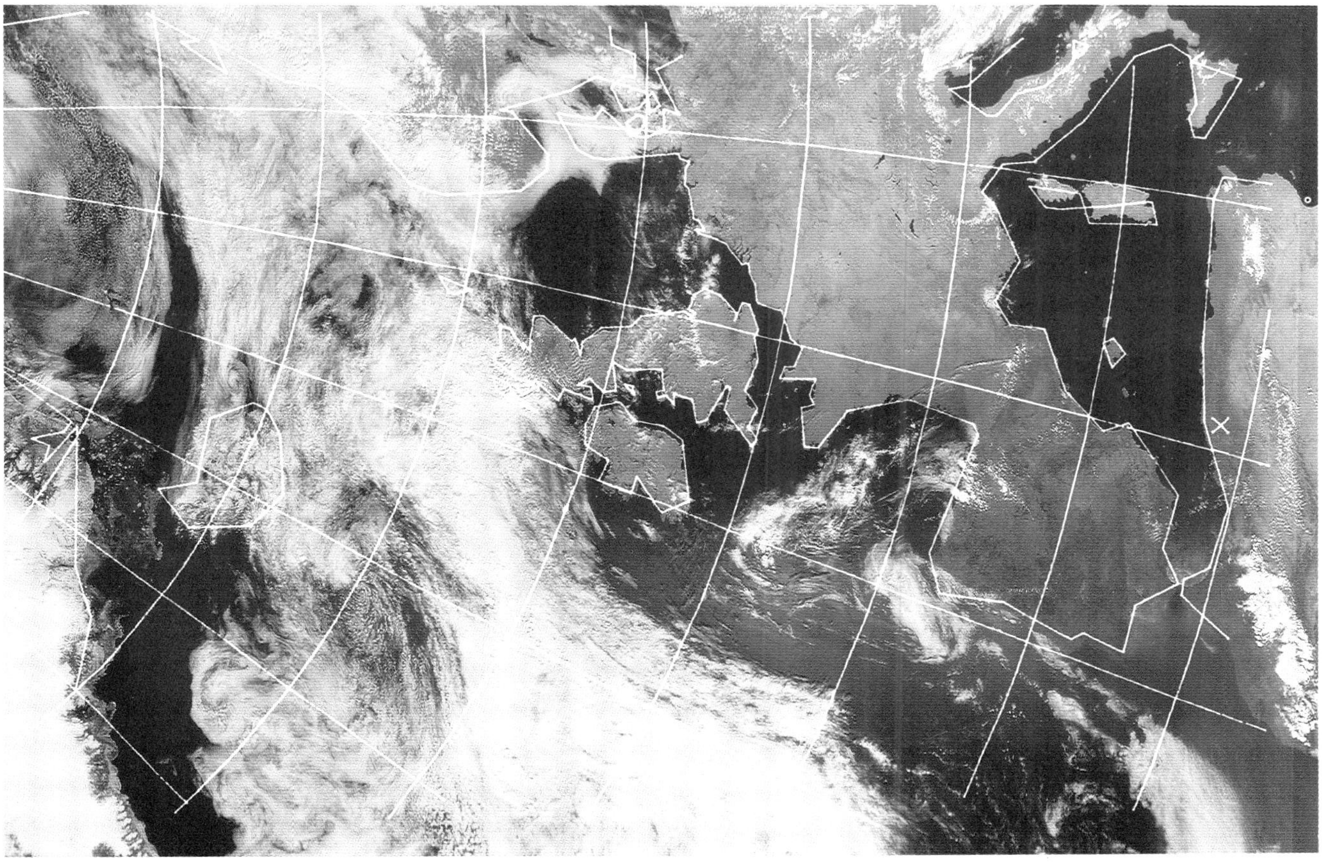

Image 1 – visible, 7 July 1984, 15.18 © University of Dundee

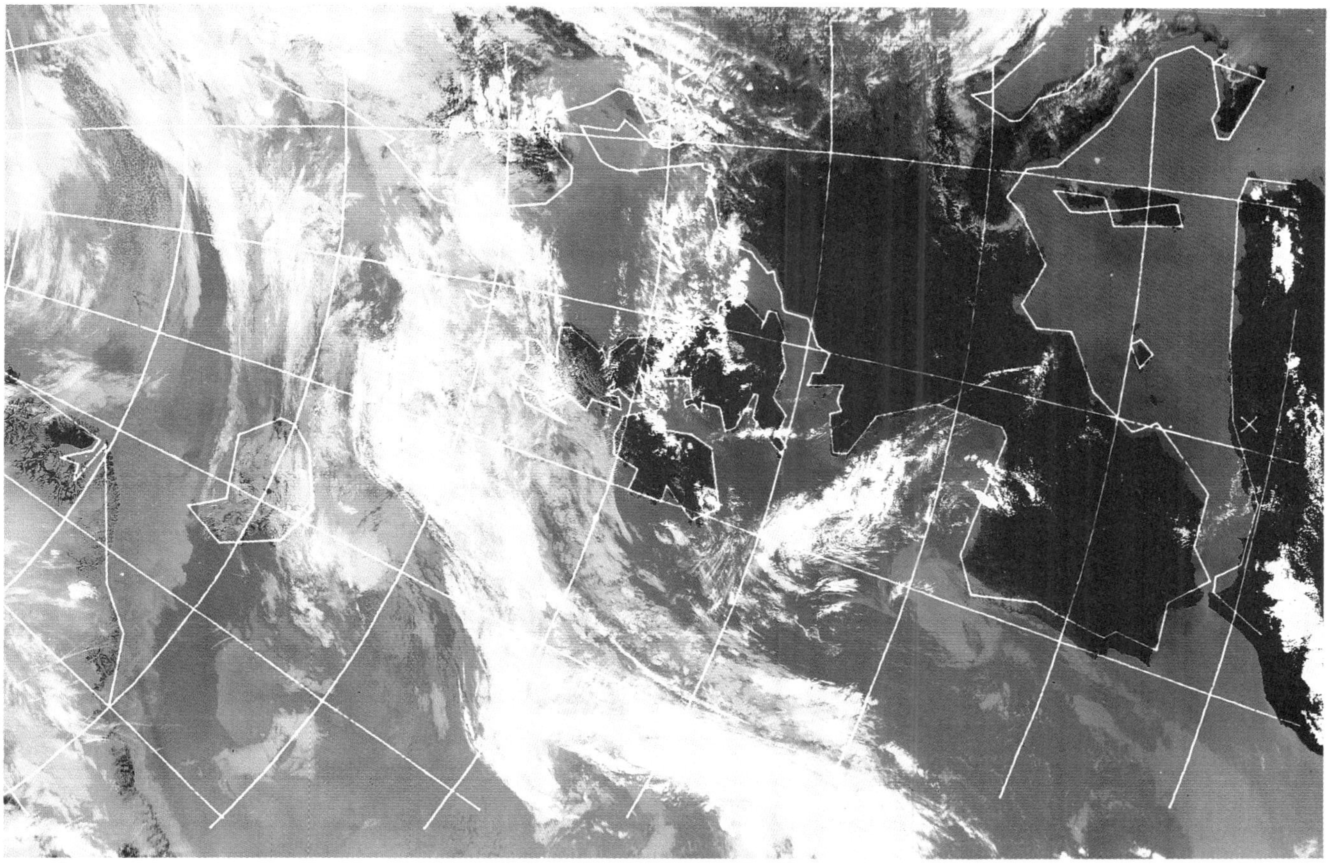

Image 2 – infra-red, 7 July 1984, 15.18 © University of Dundee

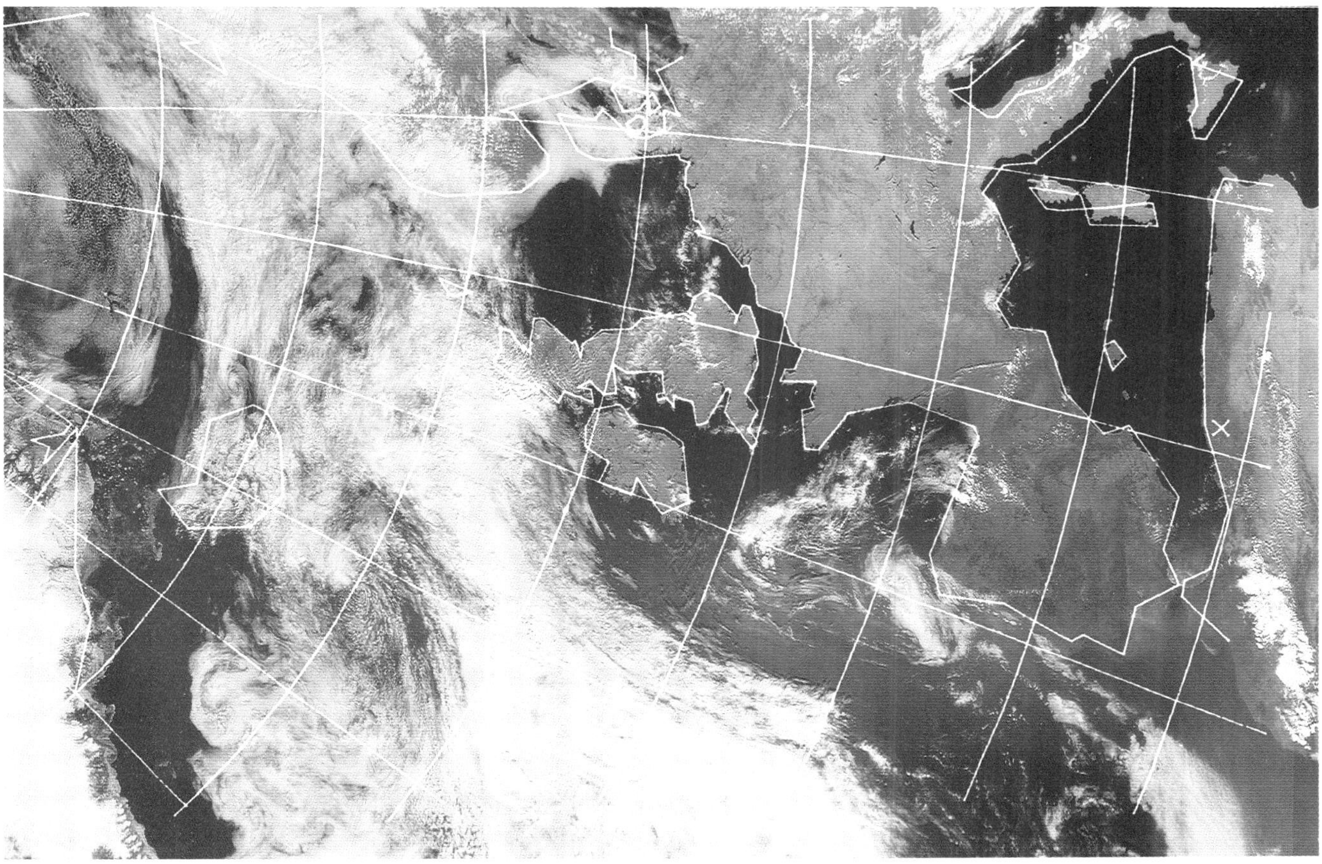

Image 1 – visible, 7 July 1984, 15.18 © University of Dundee

Image 4 – infra-red, 1 August 1979, 14.23 © University of Dundee

Image 3 – visible, 1 August 1979, 14.23 © University of Dundee

Image 4 – infra-red, 1 August 1979, 14.23 © *University of Dundee*

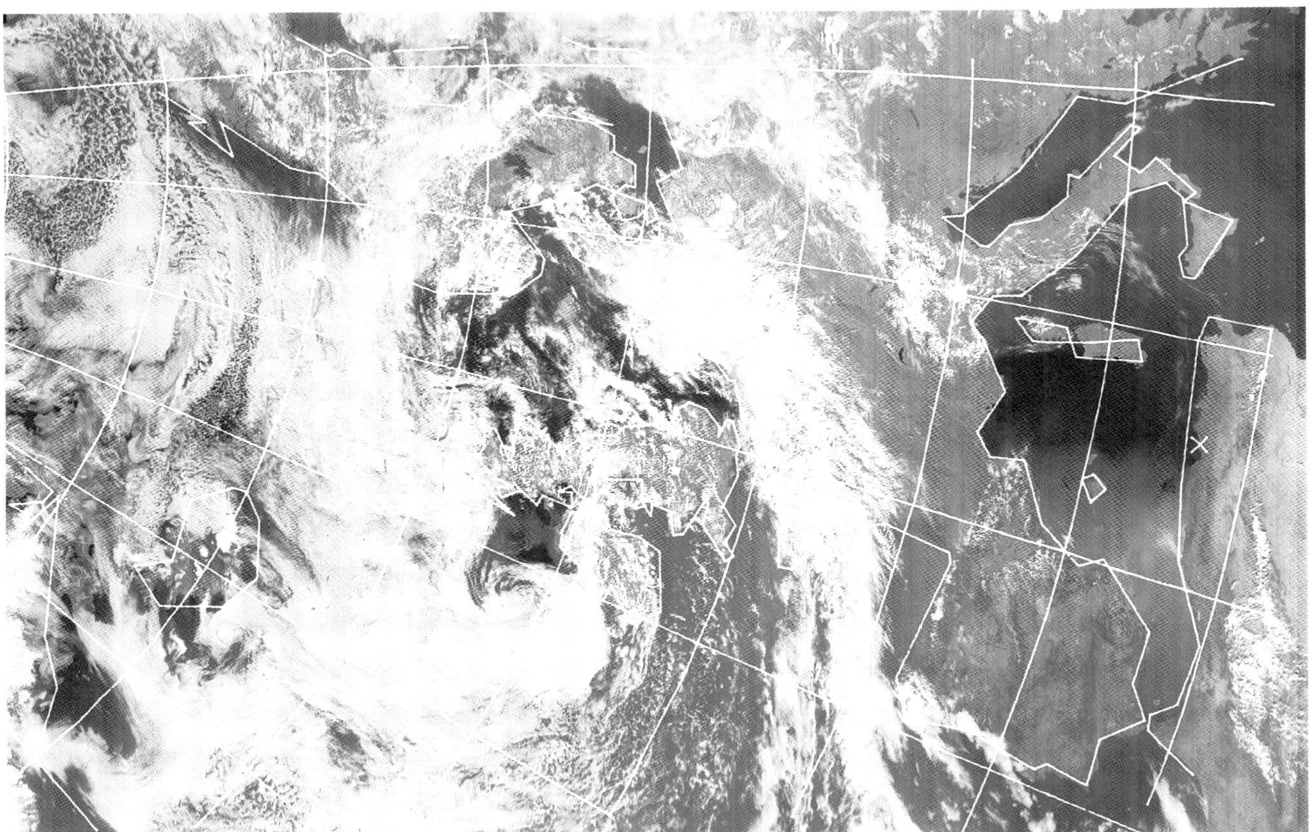

Image 3 – visible, 1 August 1979, 14.23 © *University of Dundee*

Image 4 – infra-red, 1 August 1979, 14.23 © University of Dundee

Image 3 – visible, 1 August 1979, 14.23 © University of Dundee

Image 4 – infra-red, 1 August 1979, 14.23 © University of Dundee

Image 3 – visible, 1 August 1979, 14.23 © University of Dundee

Image 4 – infra-red, 1 August 1979, 14.23 © University of Dundee

Image 3 – visible, 1 August 1979, 14.23 © University of Dundee

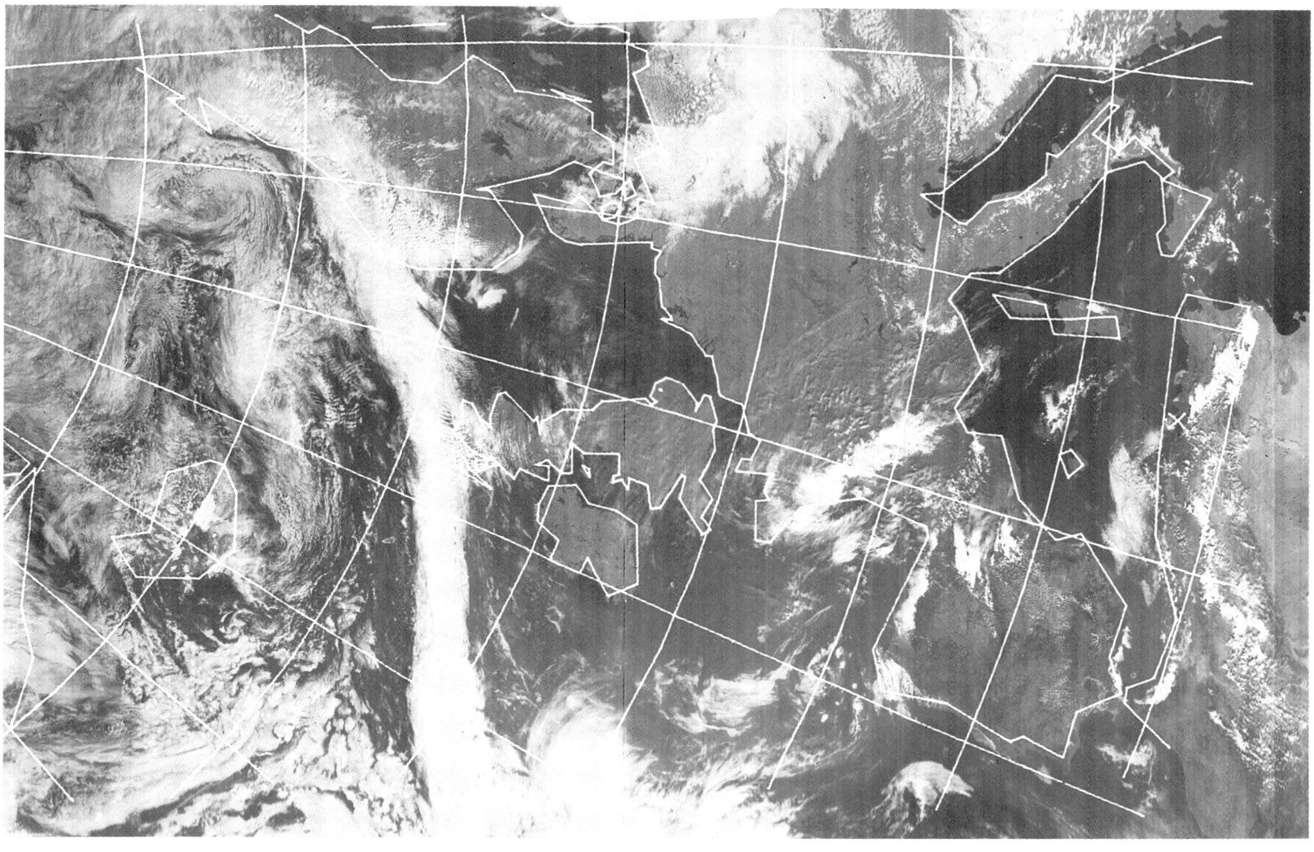

Image 6 – visible, 29 August 1979, 14.27 © University of Dundee

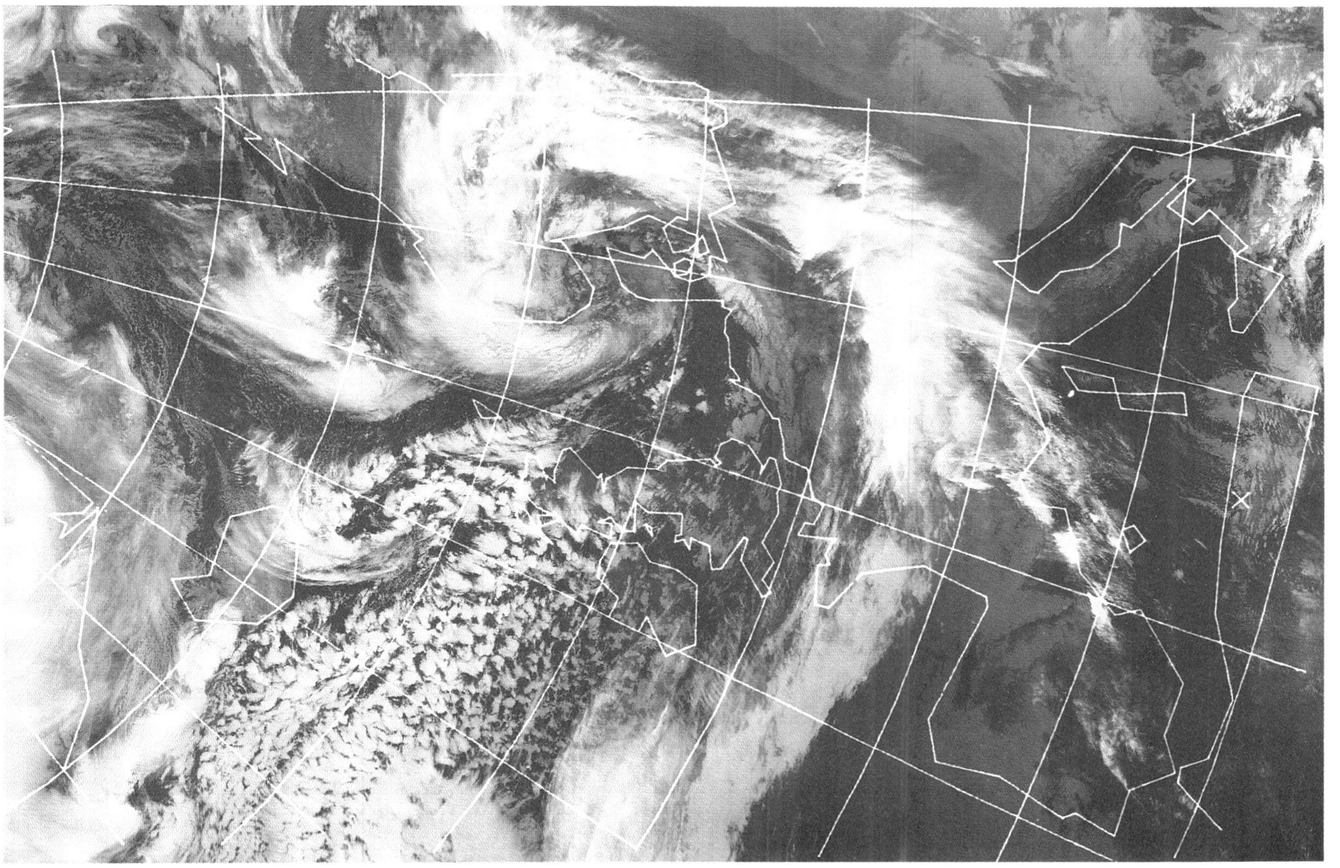

Image 5 – infra-red, 13 January 1984, 14.30 © University of Dundee

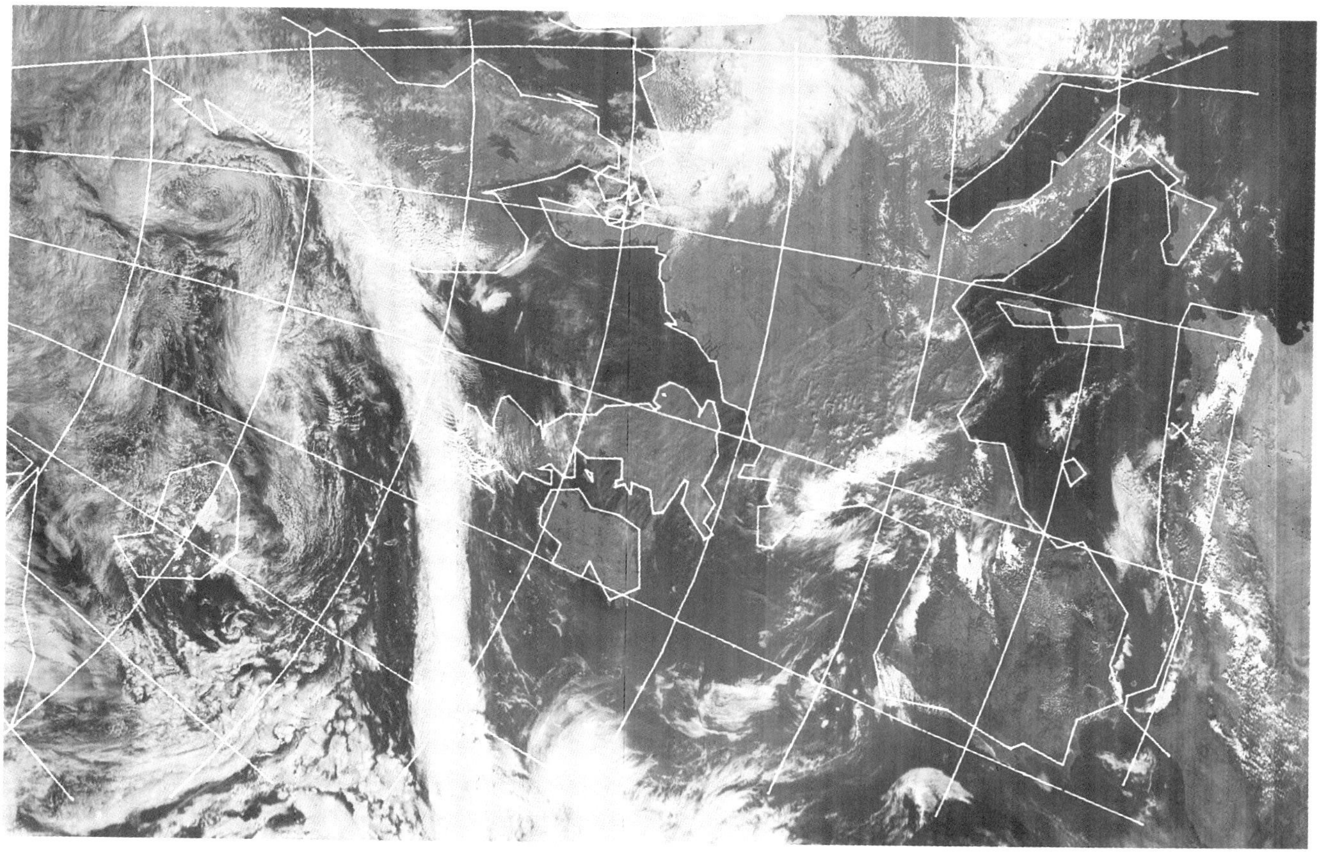

Image 6 – visible, 29 August 1979, 14.27 © University of Dundee

Image 5 – infra-red, 13 January 1984, 14.30 © University of Dundee

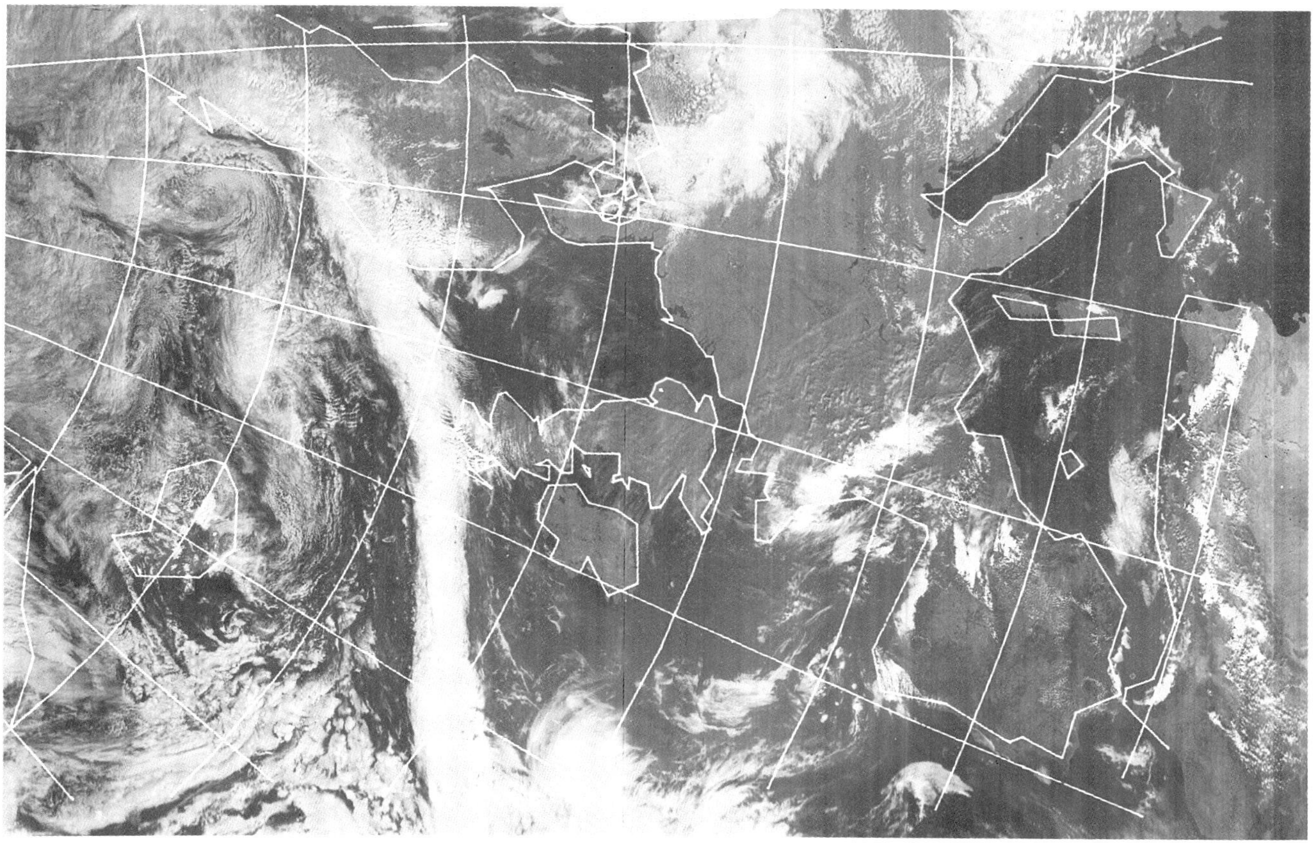

Image 6 – visible, 29 August 1979, 14.27 © University of Dundee

Image 5 – infra-red, 13 January 1984, 14.30 © University of Dundee

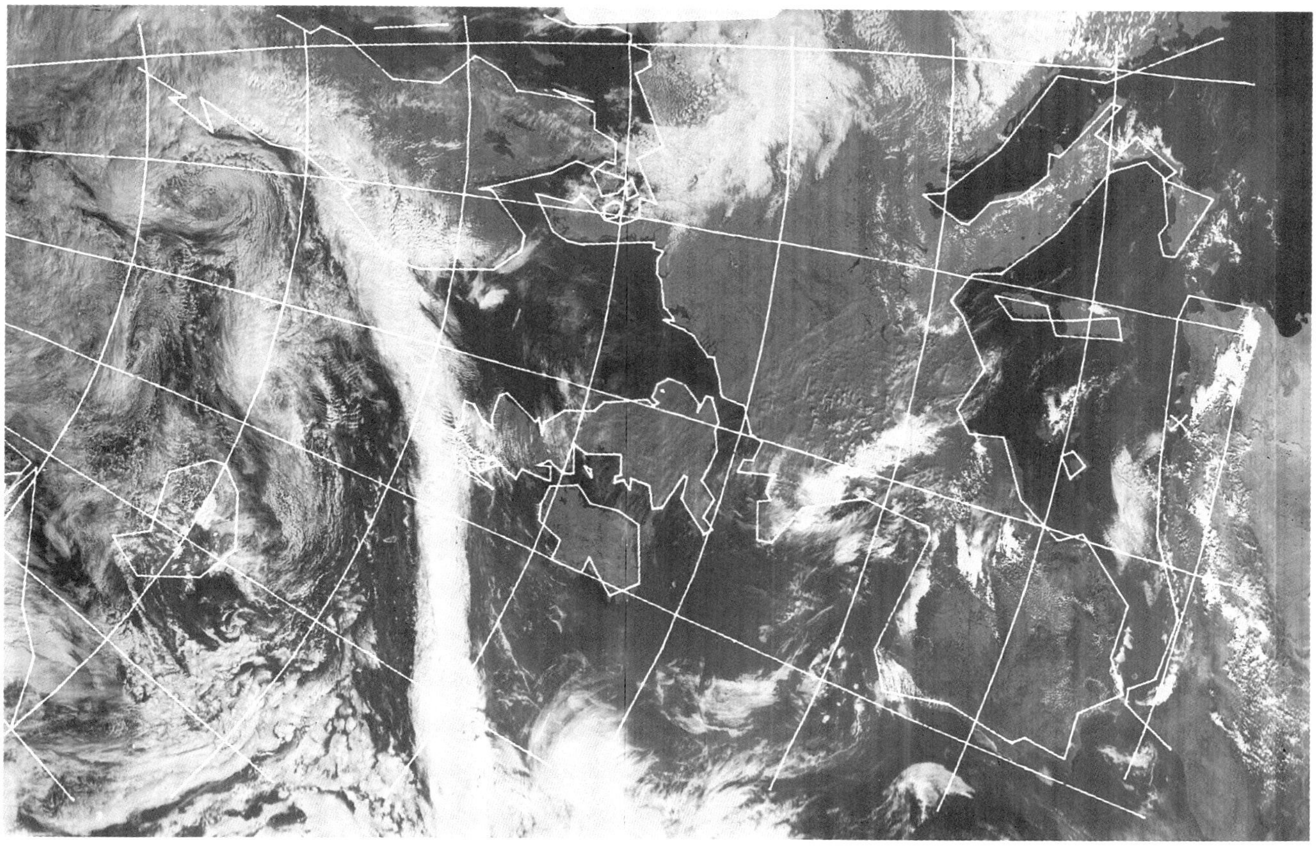

Image 6 – visible, 29 August 1979, 14.27 © University of Dundee

Image 5 – infra-red, 13 January 1984, 14.30 © University of Dundee

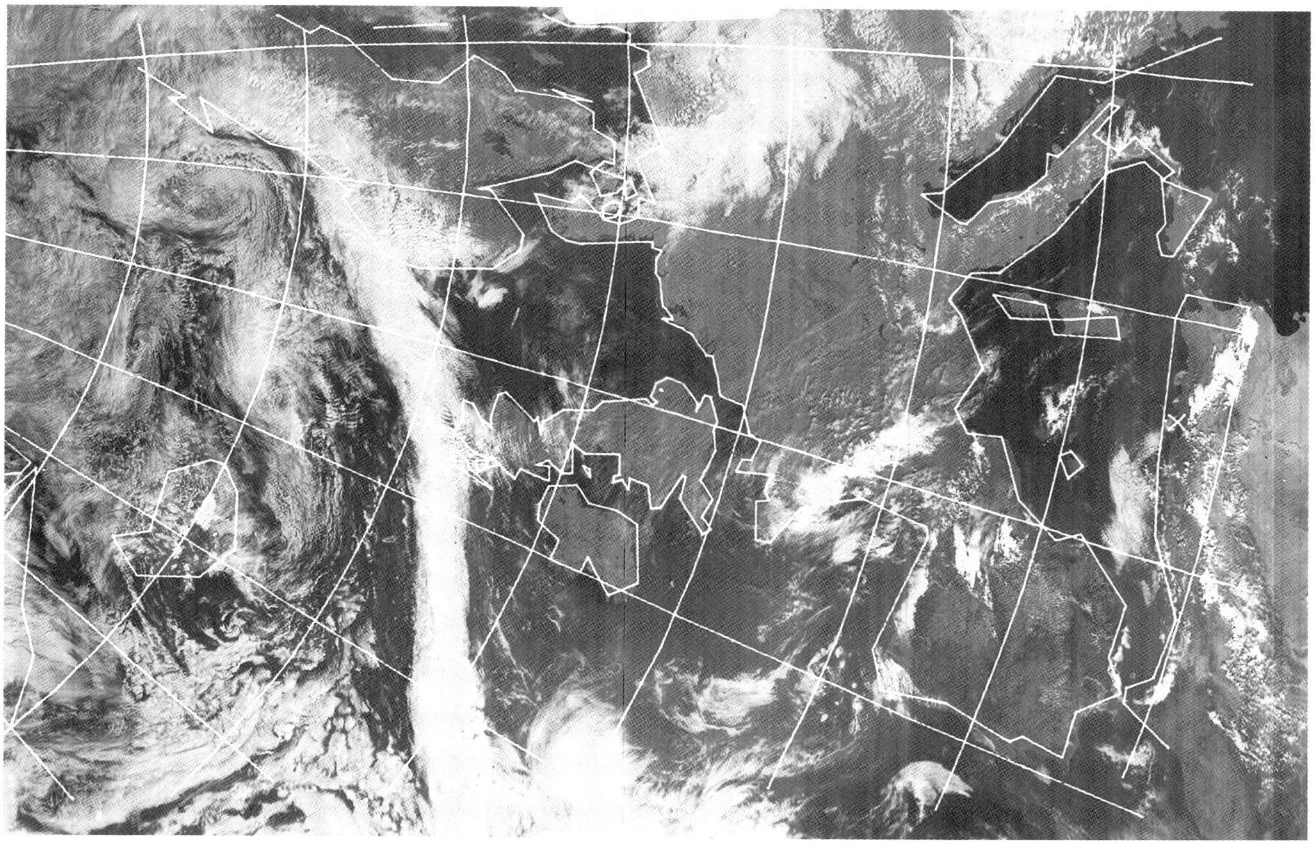

Image 6 – visible, 29 August 1979, 14.27 © University of Dundee

Image 5 – infra-red, 13 January 1984, 14.30 © University of Dundee

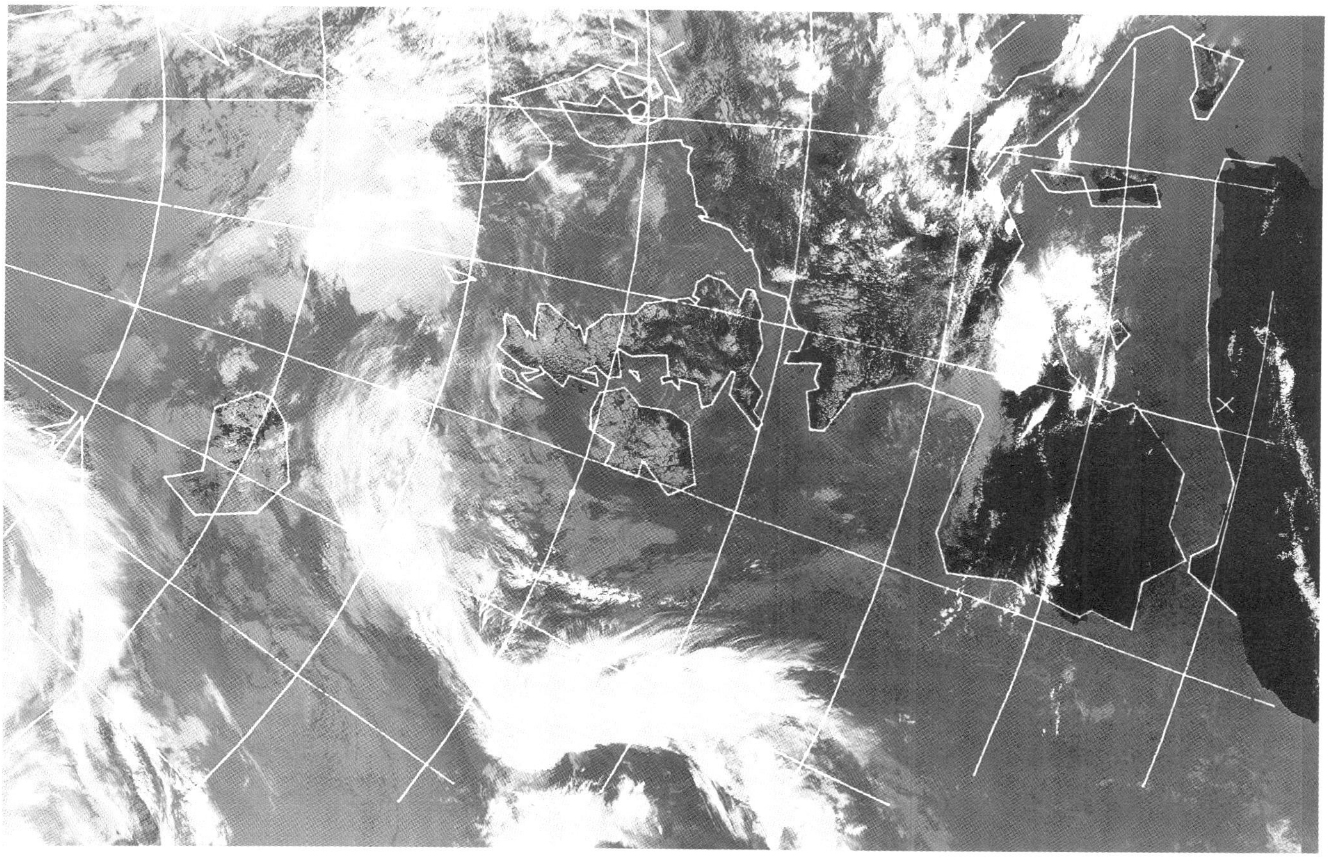

Image 8 – infra-red, 16 August 1984, 15.27 © University of Dundee

Image 7 – infra-red, 14 January 1984, 14.17 © University of Dundee

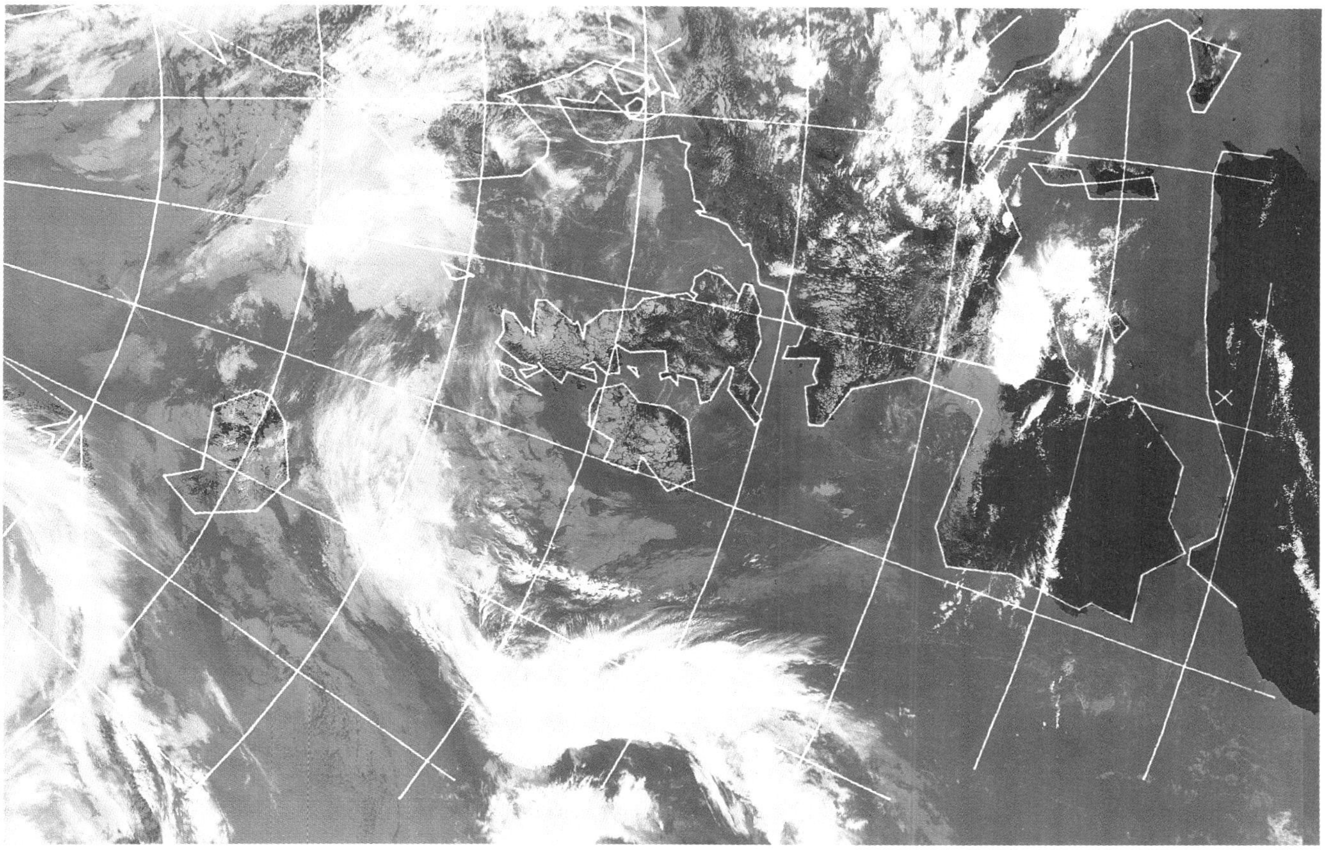

Image 8 – infra-red, 16 August 1984, 15.27 © University of Dundee

Image 7 – infra-red, 14 January 1984, 14.17 © University of Dundee

Image 8 – infra-red, 16 August 1984, 15.27 © University of Dundee

Image 7 – infra-red, 14 January 1984, 14.17 © University of Dundee

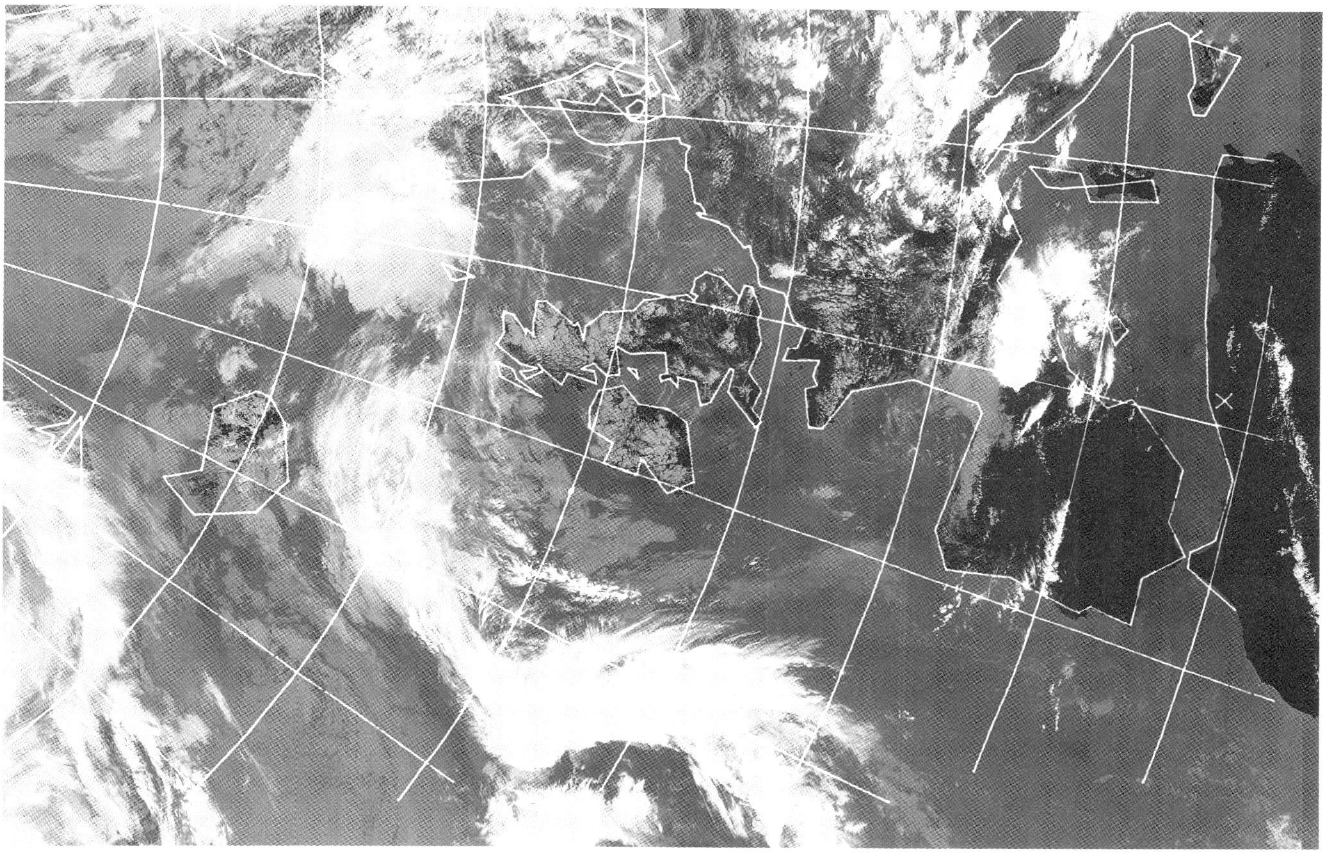

Image 8 – infra-red, 16 August 1984, 15.27 © University of Dundee

Image 7 – infra-red, 14 January 1984, 14.17 © University of Dundee

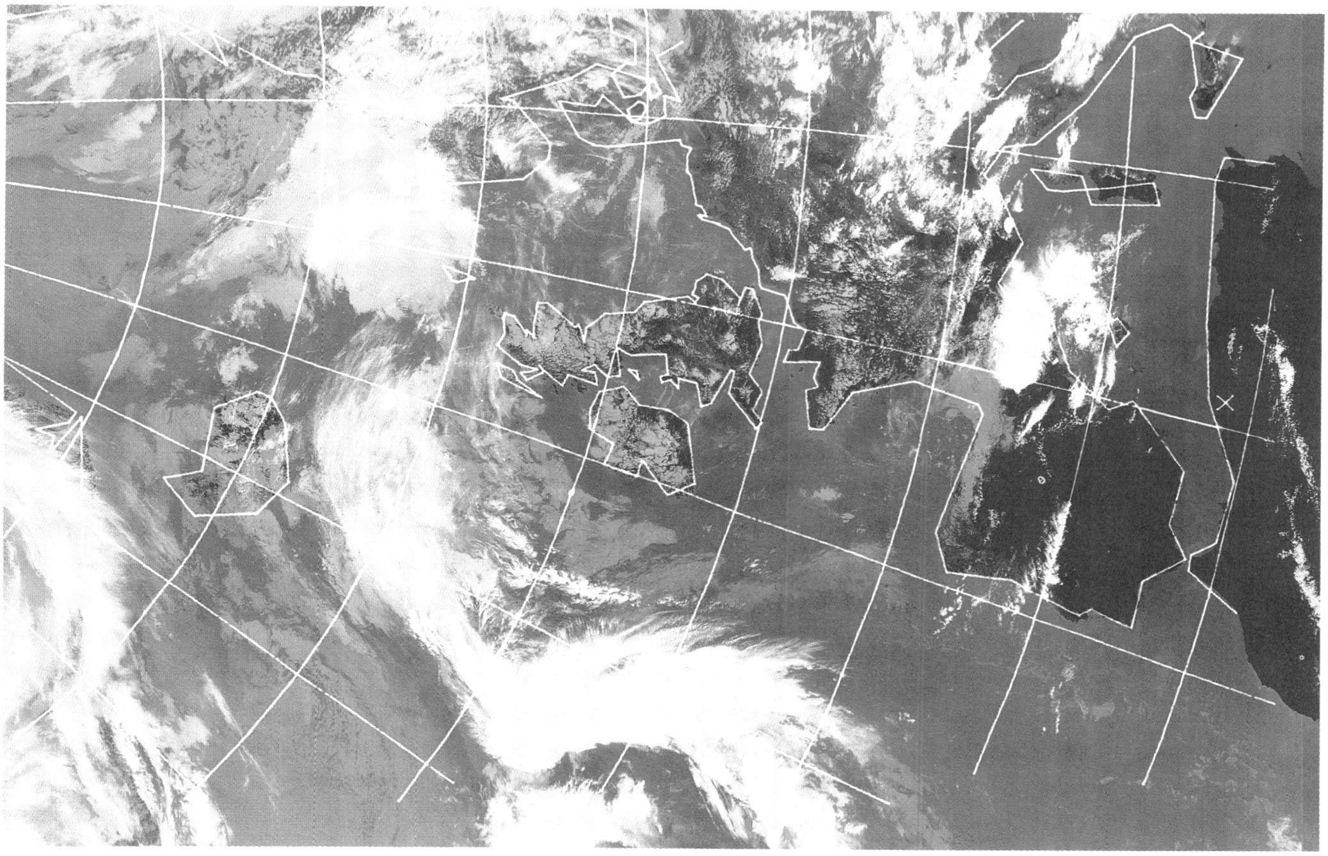

Image 8 – infra-red, 16 August 1984, 15.27 © University of Dundee

Image 7 – infra-red, 14 January 1984, 14.17 © University of Dundee

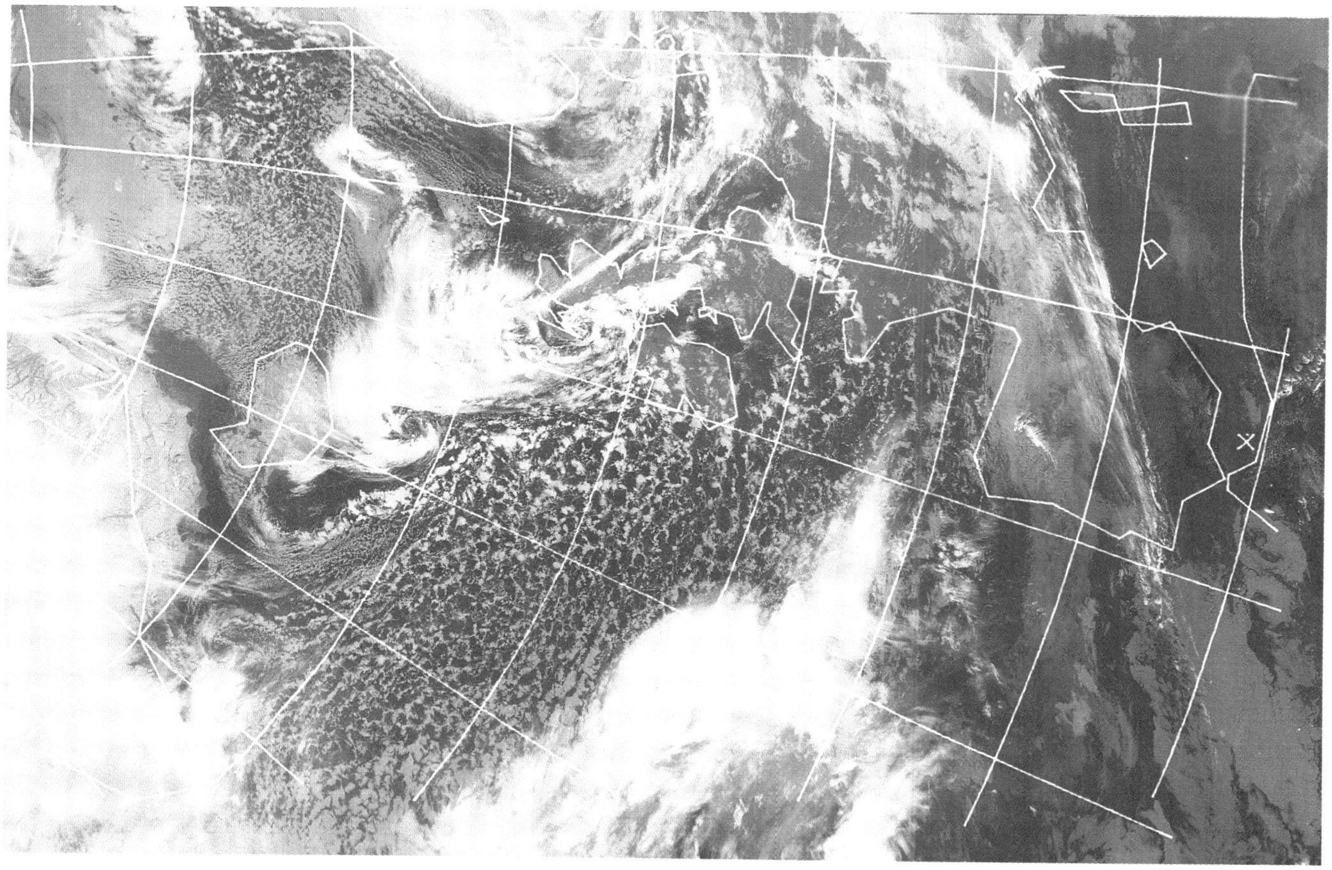

Image 10 – infra-red, 17 January 1984, 13.53 © University of Dundee

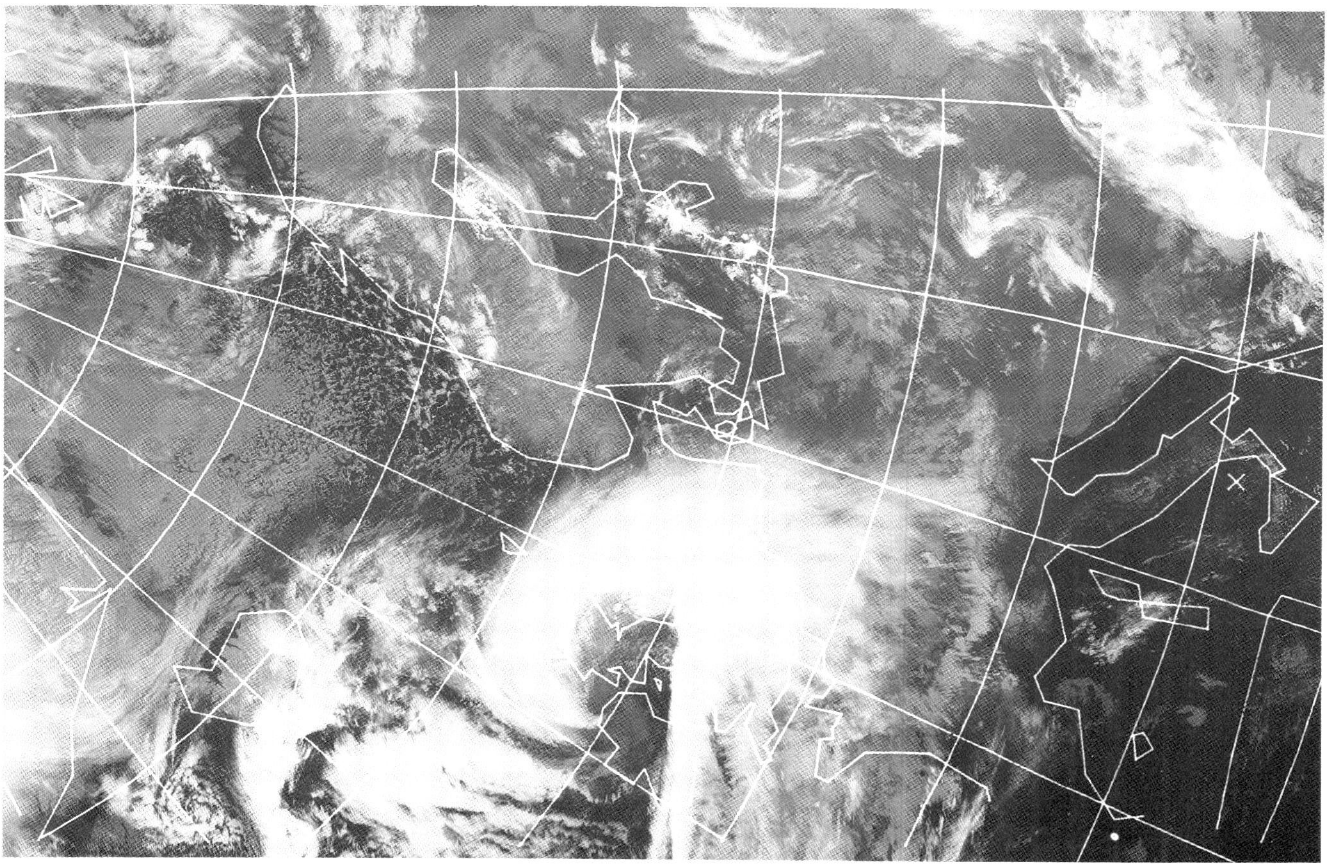

Image 9 – infra-red, 16 January 1984, 13.53 © University of Dundee

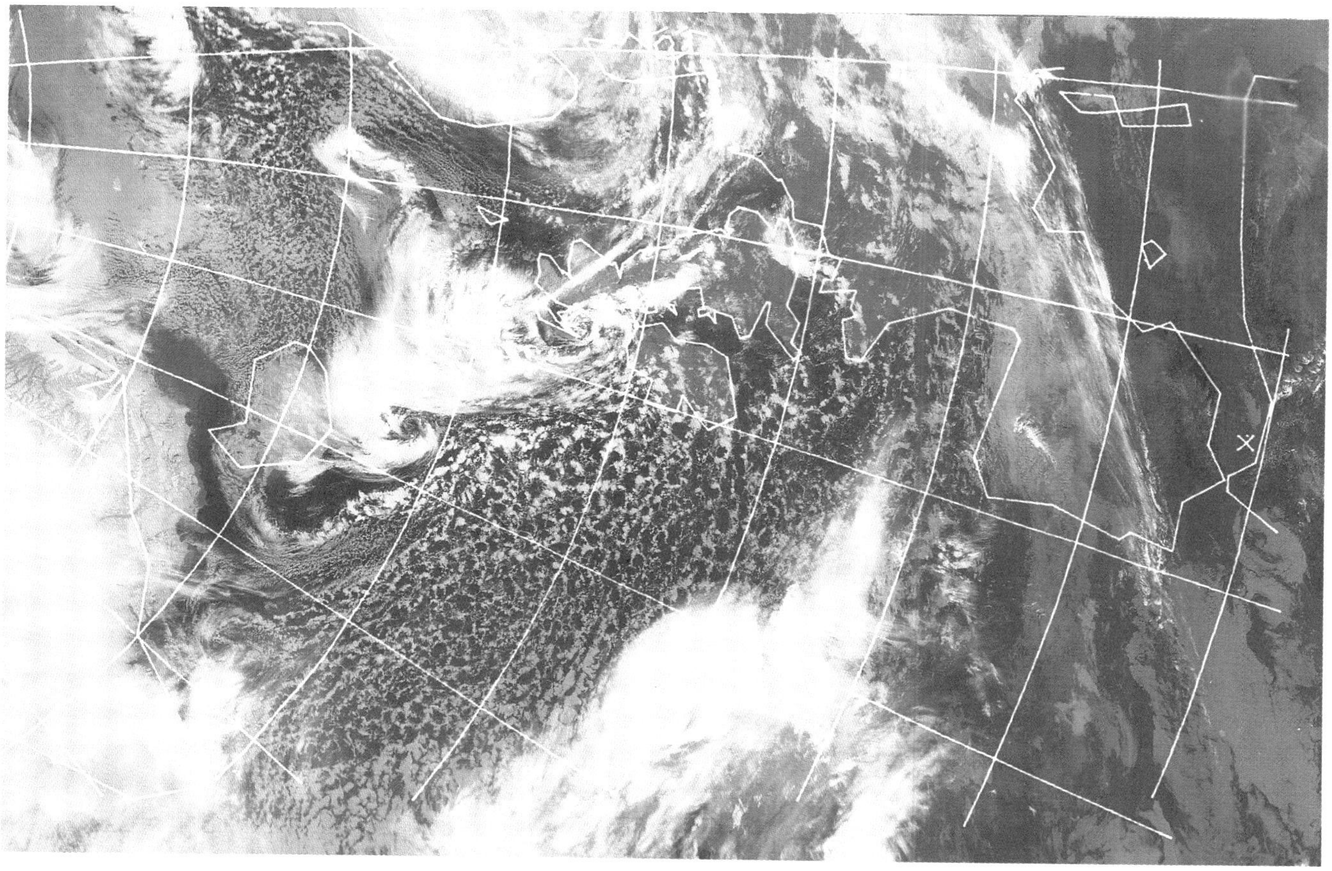

Image 10 – infra-red, 17 January 1984, 13.53 © University of Dundee

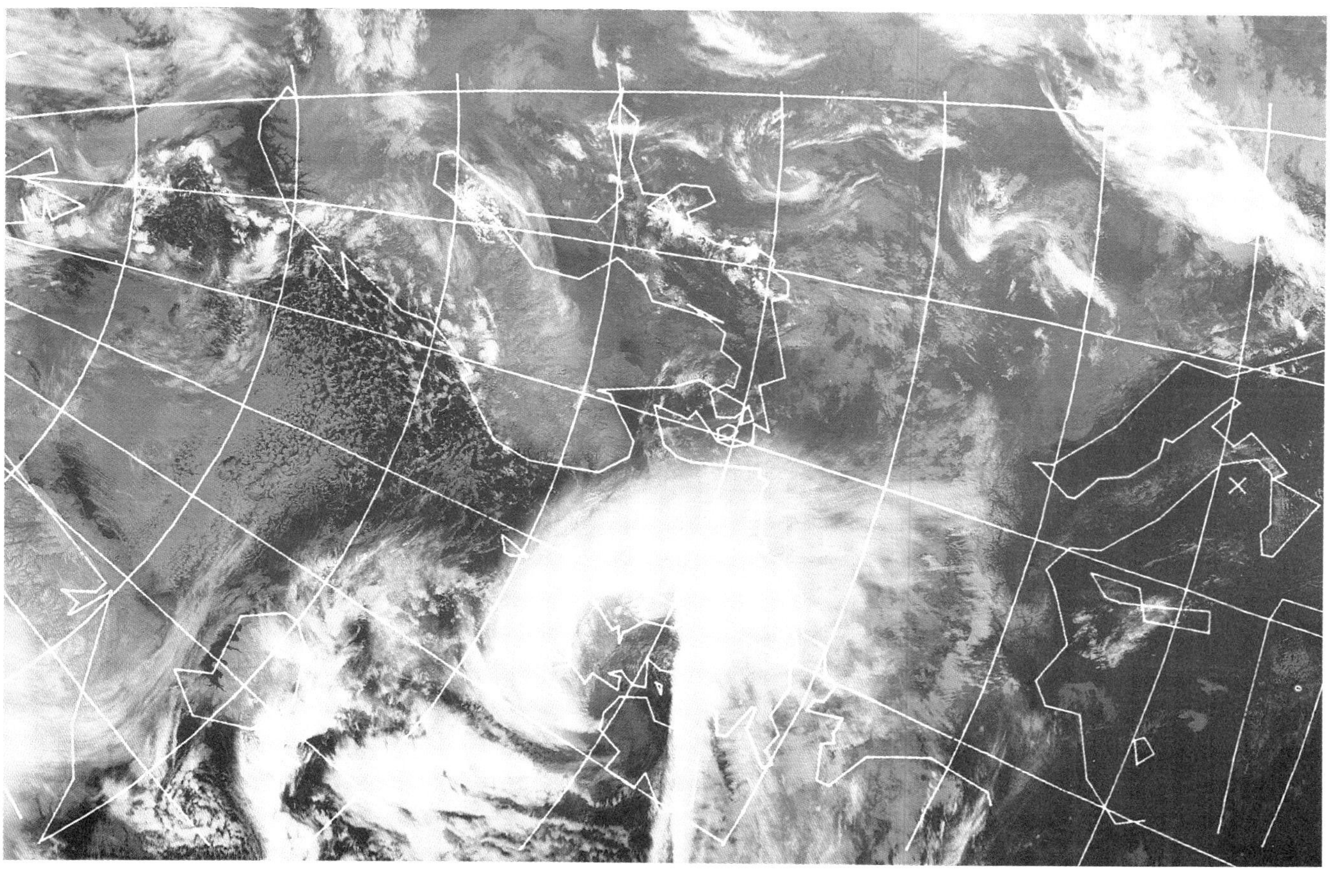

Image 9 – infra-red, 16 January 1984, 13.53 © University of Dundee

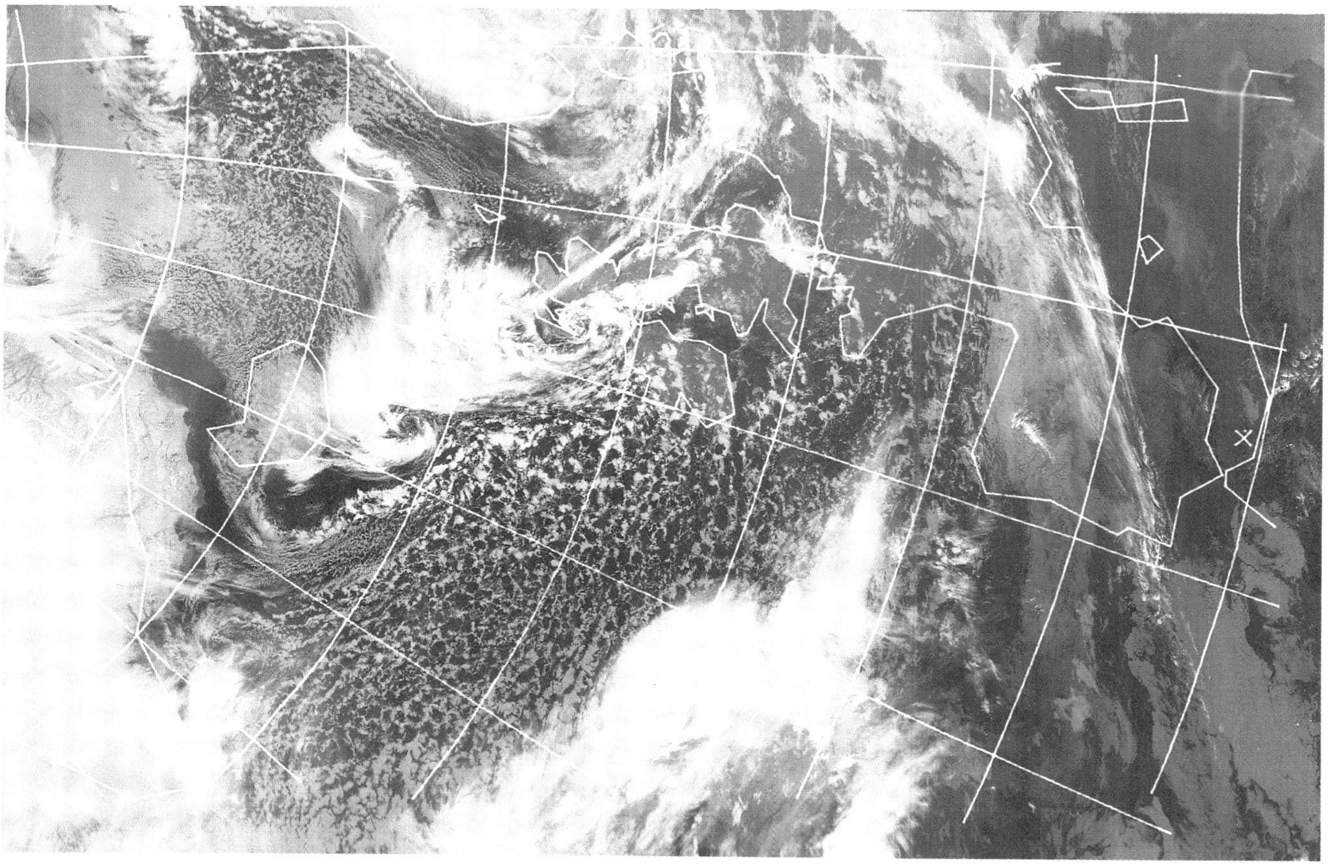

Image 10 – infra-red, 17 January 1984, 13.53 © University of Dundee

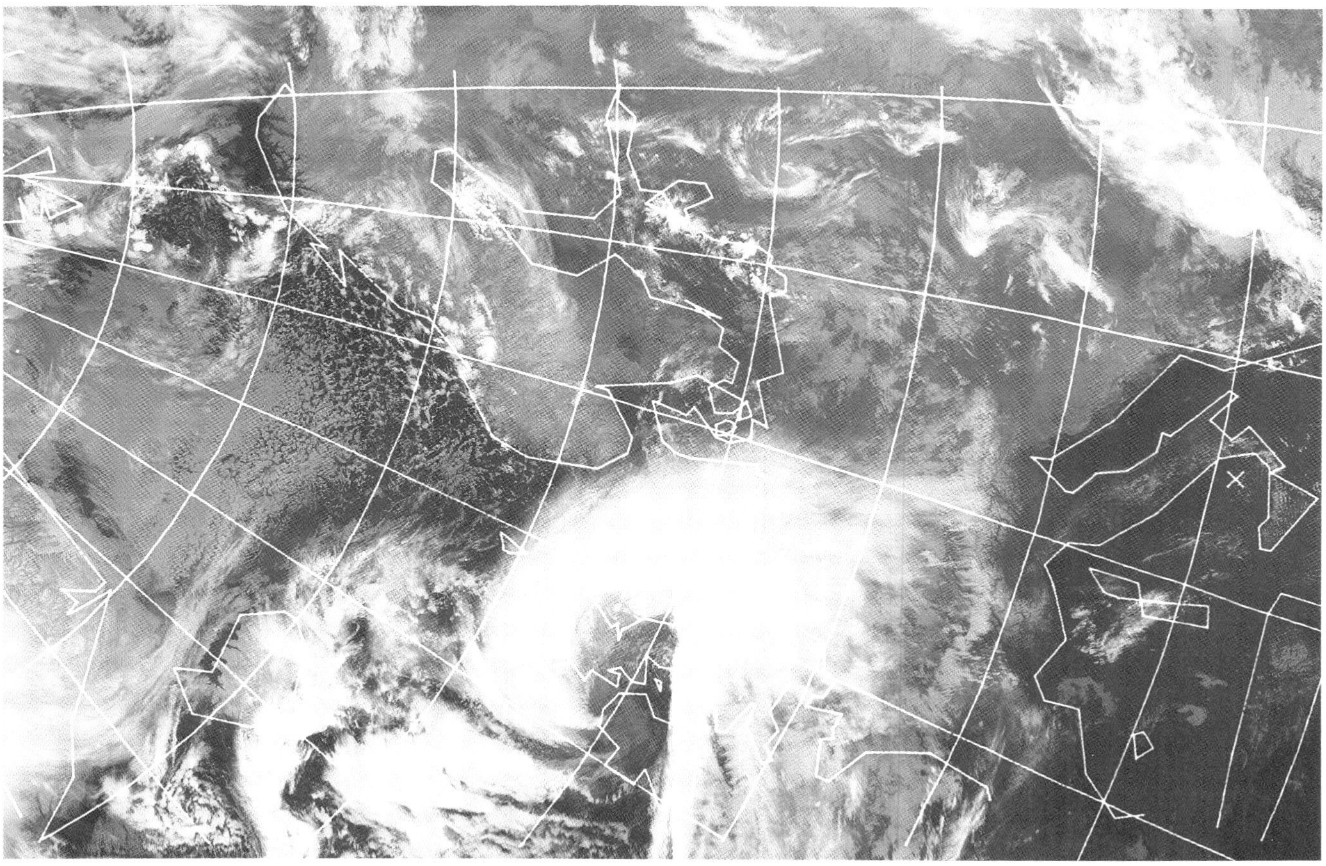

Image 9 – infra-red, 16 January 1984, 13.53 © University of Dundee

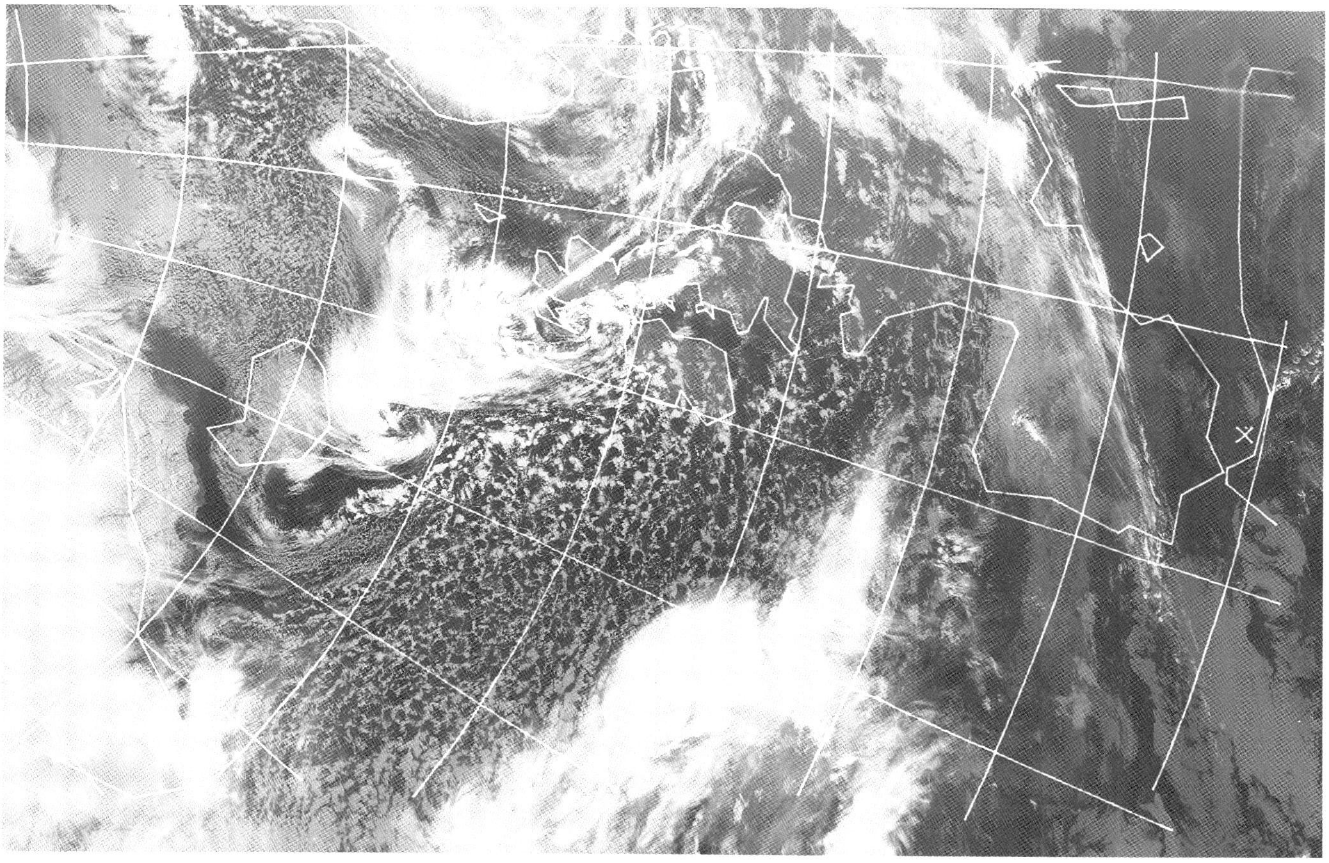

Image 10 – infra-red, 17 January 1984, 13.53 © University of Dundee

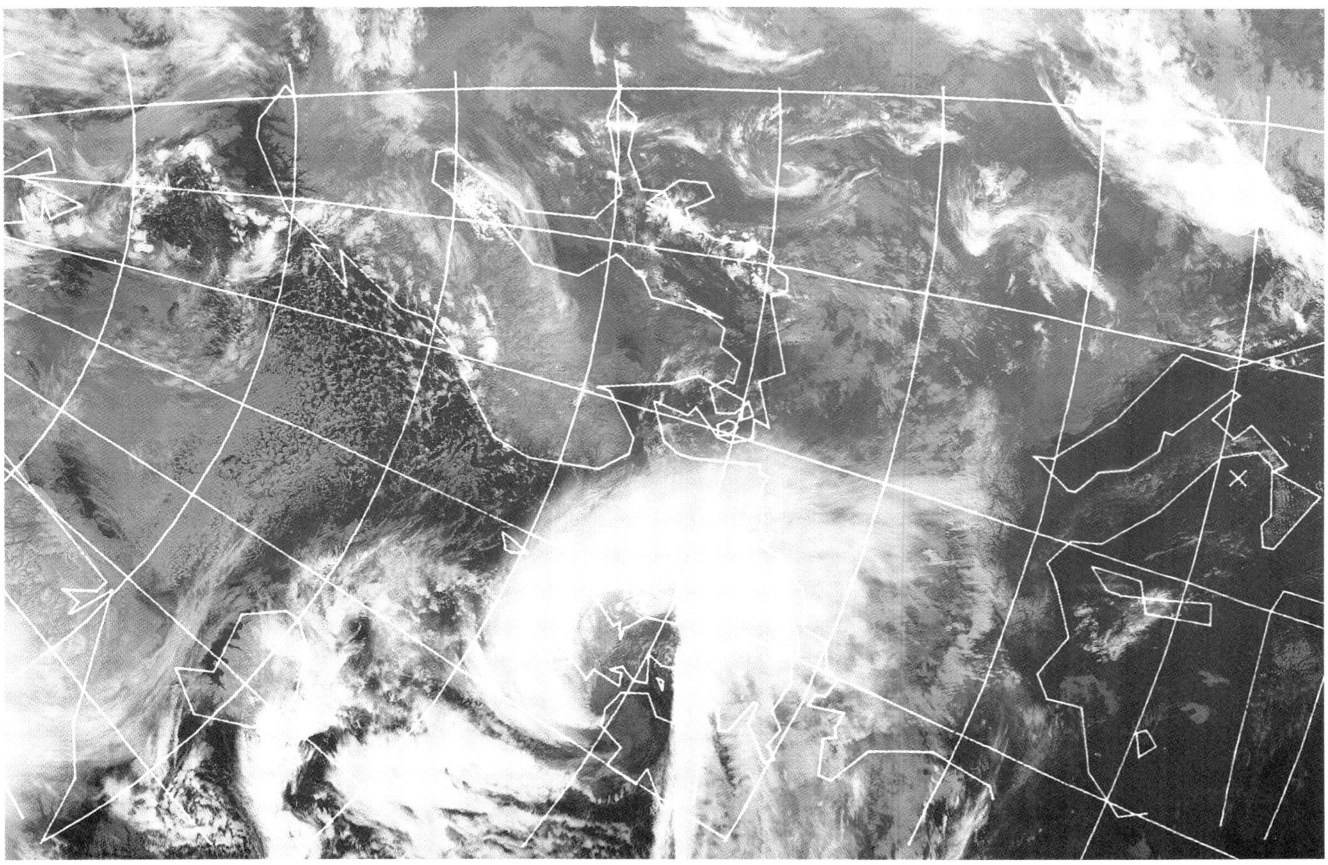

Image 9 – infra-red, 16 January 1984, 13.53 © University of Dundee

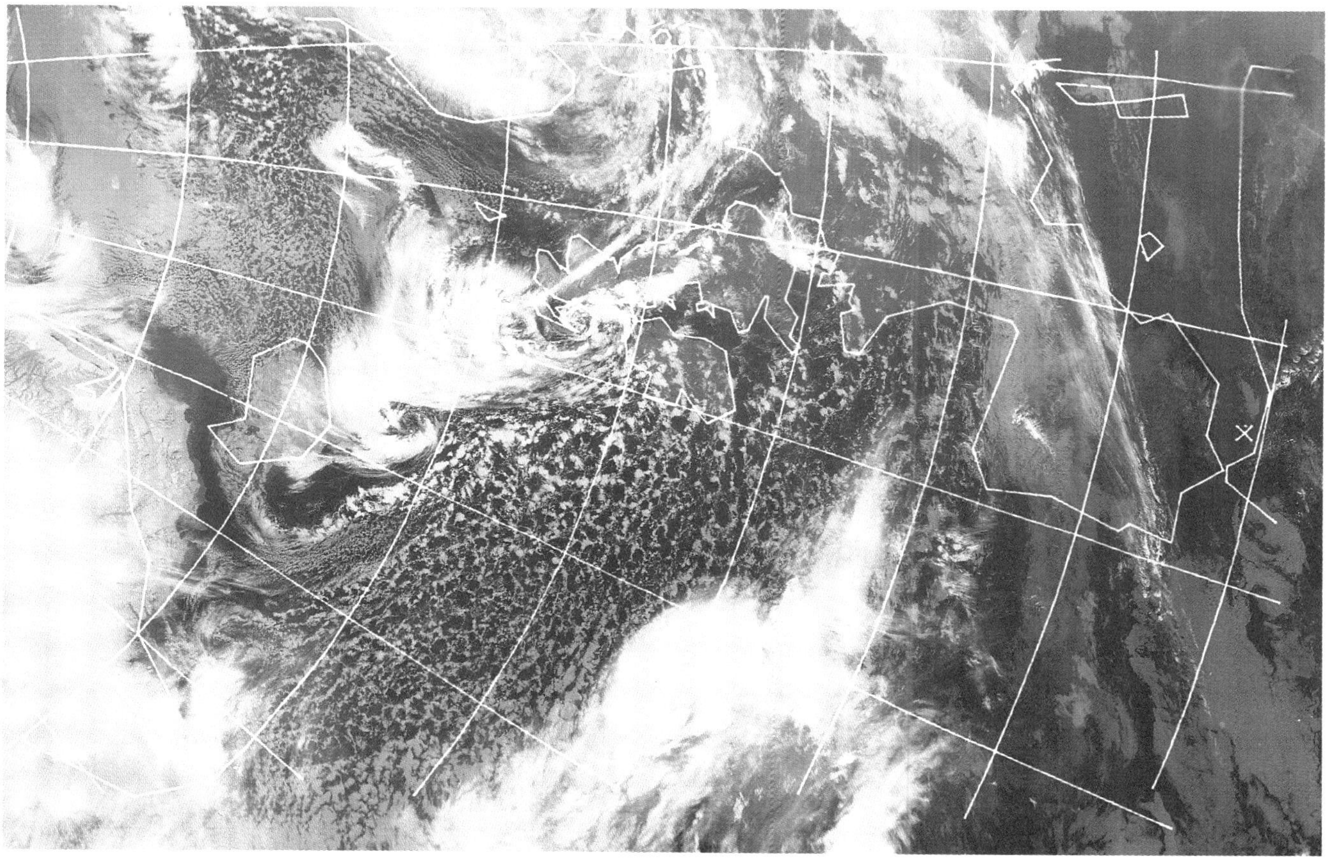

Image 10 – infra-red, 17 January 1984, 13.53 © University of Dundee

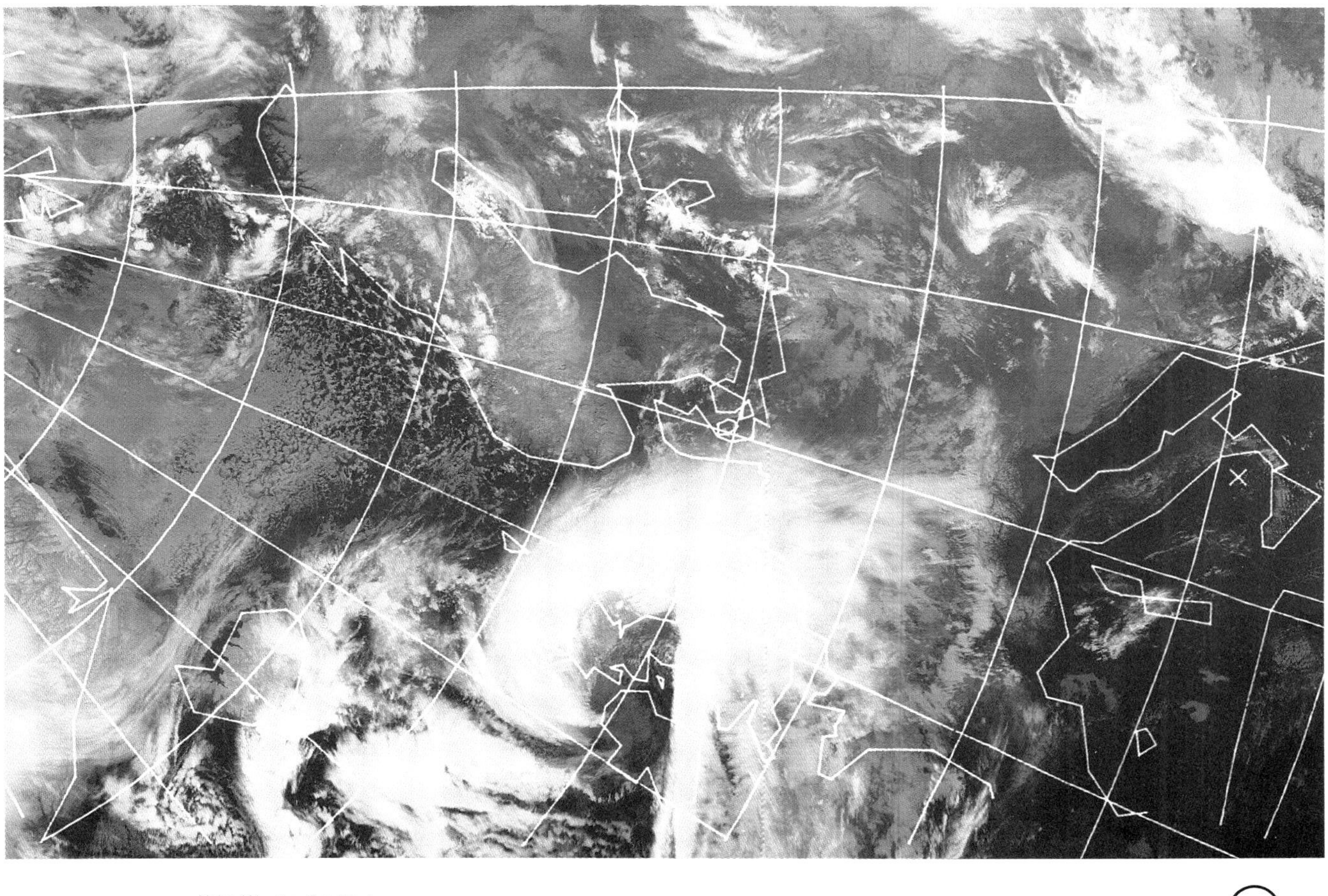

Image 9 – infra-red, 16 January 1984, 13.53 © University of Dundee

British National Space Centre

The Information Officer
British National Space Centre
Millbank Tower
Millbank
London
SW1P 4QU

Provides general satellite information.

Dartcom

Dartcom
Postbridge
Devon
PL20 6SY

Provides software for generating Meteosat timetables.

European Space Agency

Meteosat Exploitation Project
Robert Bosch Strasse 5
6100 Darmstadt
Federal Republic of Germany

Produces useful information about Meteosat, although publications tend to be very technical.

Griffin and George Ltd.

Griffin and George Ltd.
Bishop Meadow Road
Loughborough
LE11 ORG

Provides SAT FOOT, an orbital prediction program for NOAA (for BBC computers), and software for generating Meteosat timetables.

Meteorological Office

London Road
Bracknell
Berkshire
RG21 2UR

Provides a library and archives service of both weather maps/statistics and satellite images.

National Oceanographic and Atmospheric Administration

Satellite Data Service Division
World Weather Building
Room 100
Washington DC 20233
USA

Produces useful information on NOAA, including a timetable of NOAA passes.

National Remote Sensing Centre

The Education Officer
National Remote Sensing Centre
RAE Space Department
Farnborough
Hampshire
GU14 6TD

Produces useful educational resources on satellites.

Pictorial charts

Pictorial Charts Educational Trust
27 Kirchen Road
London
W13 OUD

Provides a number of useful poster sets on satellites and weather.

Royal Meteorological Society

James Glaisher House
Grenville Place
Bracknell
Berkshire
RG12 1BX

Provides a number of useful teaching resources such as SAT PAKS 1 and 2, and some slide sets.

The Slide Centre

Ilton
Ilminster
Somerset
TA19 9HS

Provides a number of useful satellite slide sets.

Unilab Ltd.

Unilab Ltd.
Clarendon Road
Blackburn
BB1 9TA

Provides SAT PAK 1, an orbital prediction program for NOAA (for BBC computers).

University of Dundee

Satellite Receiving Station
University of Dundee
Dundee
DD1 4HN

Provides high-quality NOAA satellite images.

Weatherwatch International UK

RAE Lasham
Alton
Hampshire
GU34 5SH

Provides a recorded message service on the up-to-date orbit details for NOAA – ring 025 638 448 between 17.30 and 08.30 on weekdays or any time at weekends. Distributes packs of free information about various weather satellites, and up-to-date timetables for Meteosat.

Simply colour these symbols using the key provided – they can be stuck on to a large weather map using a small piece of blu-tack on each one.

Key 1 black 2 white 3 yellow 4 red 5 blue

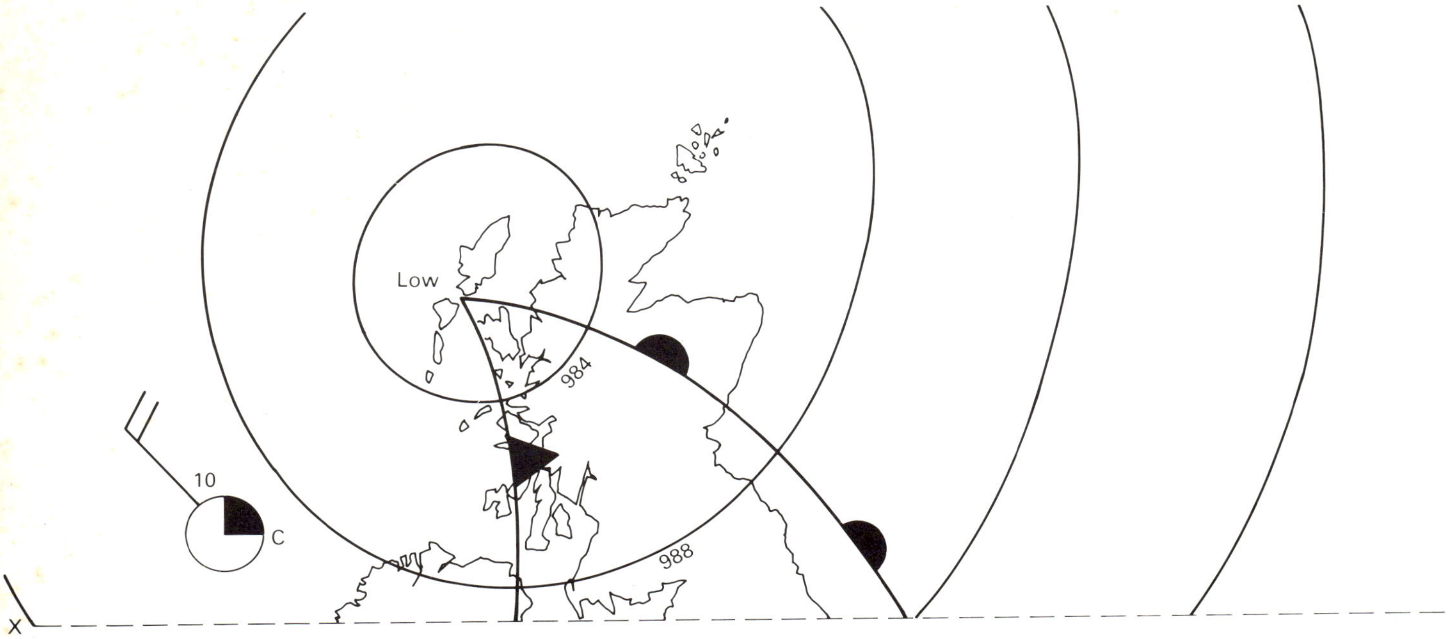

Fold along the dashed lines X, Y and Z, and then fold up along the centre dashed line Y.

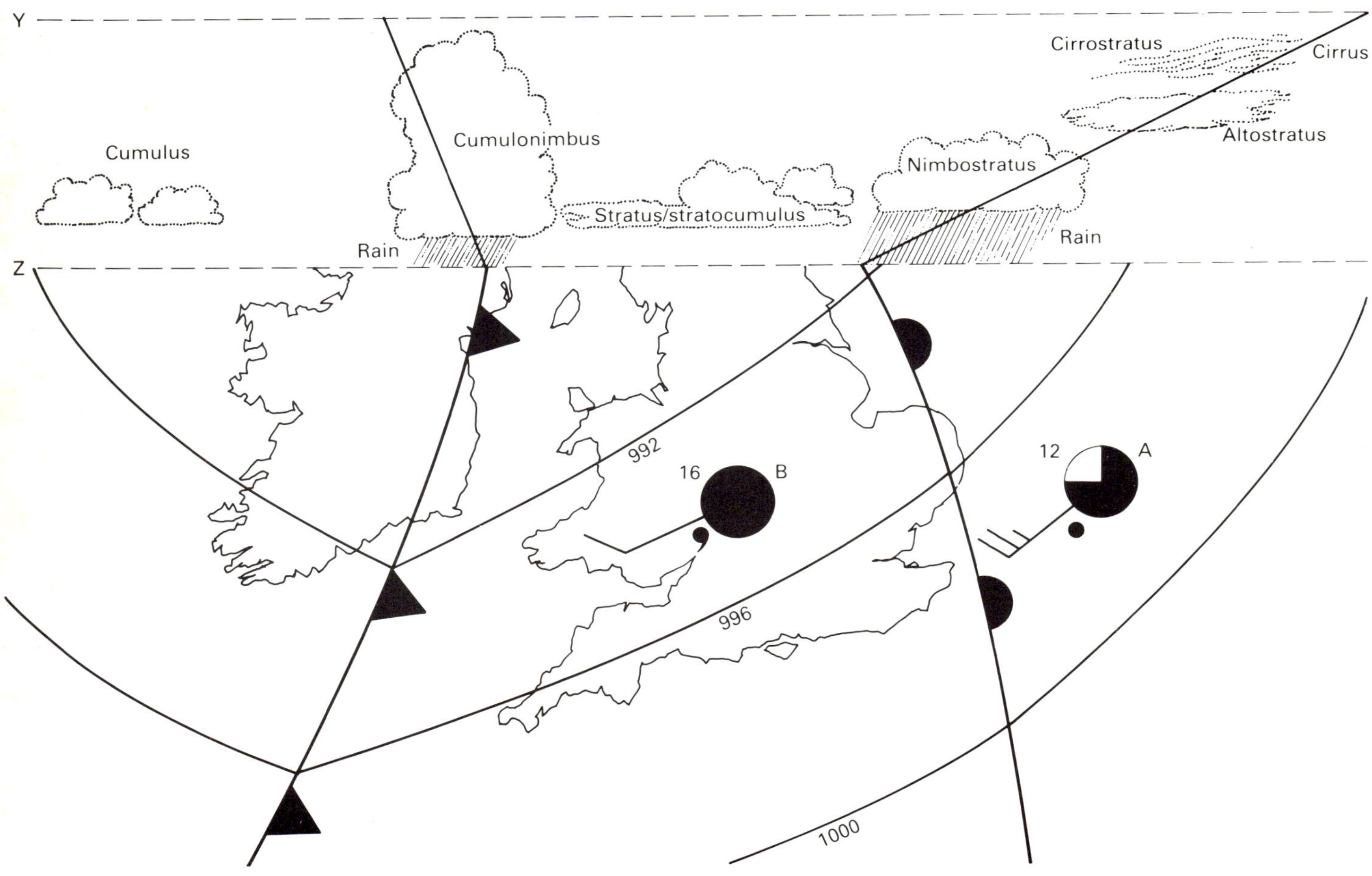